The New Celebrity Scientists

THE
New CELEBRITY
SCIENTISTS

Out of the Lab and into the Limelight

DECLAN FAHY

ROWMAN & LITTLEFIELD
Lanham • Boulder • New York • London

Published by Rowman & Littlefield
A wholly owned subsidiary of The Rowman & Littlefield Publishing Group,
Inc.
4501 Forbes Boulevard, Suite 200, Lanham, Maryland 20706
www.rowman.com

Unit A, Whitacre Mews, 26-34 Stannary Street, London SE11 4AB, United
Kingdom

British Library Cataloguing in Publication Information Available

Library of Congress Cataloging-in-Publication Data
Fahy, Declan.
  The new celebrity scientists : out of the lab and into the limelight / Declan
Fahy.
      pages cm
  Includes bibliographical references and index.
  ISBN 978-1-4422-3342-3 (cloth : alk. paper) — ISBN 978-1-4422-3343-0
(electronic) 1. Scientists—Biography. I. Title.
  Q141.F24 2015
  509.2'2—dc23
                                    2014039192

Printed in the United States of America

For Louisa, with love

Contents

Acknowledgments

I would like to thank Matthew Nisbet, Helena Sheehan, Brian Trench, Rodger Streitmatter, W. Joseph Campbell, Leonard Steinhorn, Kathryn Montgomery, Laura DeNardis, Amy Eisman, Caty Borum Chattoo, Massimiano Bucchi, Michael Cronin, and Aileen Fyfe for their generous advice and guidance at various stages of this book's development. I also would like to thank, for their research assistance, Jan Boyles, Conor Fahy, Amanda Winkler, Erica Sánchez-Vázquez, Kristina Brooks, Monika Thomas, and Ryan Schuette. Not least, I would like to express my sincere thanks to Leanne Silverman, Andrea O. Kendrick, and Jehanne Schweitzer at Rowman & Littlefield for all their editorial input and support.

ONE

A Brief History of Scientific Celebrity

Charles Darwin was a master of public relations. While *The Origin of Species* swept through nineteenth-century culture, shattering the prevailing religious orthodoxy about the dawn of human life, Darwin cultivated his popular image. He distributed mass-produced photographs, signed autographs, collected songs and poems about himself, responded to his voluminous mail with preprinted cards, refused most interviews, and avoided speaking before crowds where questions might catch him off guard. He met author George Eliot and received an inscribed copy of *Das Kapital* from Karl Marx.

The theory of evolution by natural selection he outlined in his 1859 book coursed through Victorian society, spreading far beyond learned journals and professional societies. The general public could buy a low-cost edition of *The Origin of Species*. Magazines spread his image and ideas through popular culture, sometimes in strange ways. Caricaturists and cartoonists, for example, drew Darwin's head—with its distinctive long beard and large dome of his skull—on the body of an ape. Darwin became public property. "Deep down," wrote one historian, "Darwin's sons and daughters were forced to accept that he was not just their father. He belonged to everybody."[1] In his time he was a scientific celebrity.

The great naturalist showed that fame, lasting fame, is never just the inevitable consequence of great achievement, even one as seismic as *The Origin of Species*. The world must hear about the achievement.

Nowhere is this more clear than in the case of the twentieth century's most iconic scientist—physicist Albert Einstein. He personified science in mass culture, and he became a global symbol for the mind's phenomenal power. He exploded into popular consciousness in 1919 after his general theory of relativity, which was nothing less than a new way of viewing the physics of the universe, was confirmed by two independent experiments during a solar eclipse. Afterward, Einstein received significant journalistic attention—"Revolution in Science. New Theory of the Universe," reported *The Times*—but he vaulted to global celebrity two years later when he visited the United States to raise money and public awareness for Zionist causes.[2]

The two-month visit in 1921 sparked a frenzy of public and media interest. The *New York Times* described a "man in a faded grey rain coat and a flopping black felt hat that nearly concealed the grey hair that straggled over his ears. . . . But underneath his shaggy locks was a scientific mind whose deductions have staggered the ablest intellects of Europe." The *New York Evening Post* described Einstein's Berlin home and detailed his love of Dostoyevsky, his working methods—lost in intense concentration in his room alone for three or four days—and his fondness for cigars.[3] The American press, wrote one scholar, was "*the* instrument that made Einstein into a celebrity."[4]

On the strength of all this coverage, as relativity matched the mood of uncertainty that followed the savagery of World War I, Einstein became a global star. The London Palladium asked him to put on his own show. Girls in Geneva mobbed him. One tried to cut off a lock of his hair. Telescopes and towers were named in his honor, as were children and cigars. When he and his wife, Elsa, attended the 1931 premiere of *City Lights*, photographers snapped their picture on the red carpet alongside the film's star, Charlie Chaplin. Abraham Pais, one of Einstein's former colleagues and a historian of science, later wrote: "Einstein, creator of some of the best science of all time, is himself a creation of the media in so far as he is and remains a public figure."[5]

As Einstein penetrated deep into popular culture in the early twentieth century, the mass media expanded dramatically. By the end of the century, the media had become a center of public life with enormous power. For most adults the media are the source of most ideas and information about science and issues affected by science. In their myriad forms, the media disseminate information; shape public opinion; and convey ideas about how the world works, how the world can be experienced, how society is organized, and what issues matter for citizens.

The media also focus overwhelmingly on individuals, leading to our pervasive celebrity culture where fame has become *the* most powerful way of understanding ideas in a complex world. Within this culture, a new type of scientist has come out of the lab and into the limelight—the celebrity scientist.

These scientific stars grip the public imagination, using their vast influence to stimulate new thinking, drive scientific controversies, enhance public understanding, mobilize social movements, and shape policy. In celebrity culture, they speak for science in public. But more than that, their fame gives them power *within* science. Their stardom affects the inner workings of science, shaping the discovery of new knowledge about our natural world.

Because of their profound ability to influence public life and professional research, the celebrity scientists constitute a new scientific elite.

FRED HOYLE AND
CARL SAGAN: MASS MEDIA STARS

In 1950, British astronomer Fred Hoyle became an emblematic figure in this new media age of science. That year, he delivered a series of BBC radio lectures on "The Nature of the Universe" that were so successful that listeners voted him Britain's most popular broadcaster. In his soft Yorkshire accent, he clearly explained cosmology and evoked familiar domestic scenes to make complex science part of listeners' everyday worlds. Even professional physicists stopped work and tuned in. The lectures were subsequently printed as *The Nature of the Universe*, which sold seventy-seven thousand copies in just six months, making it an early scientific best seller. The lecture and book made Hoyle an "international celebrity."[6]

Hoyle's rise to stardom occurred right at the beginning of a period called "the long sixties." This era spanned roughly 1955 to 1975 and it featured a dramatic clash of ideas that transformed politics, divided society, and overhauled scientific life so radically that scholars called it a second scientific revolution—after the first in the sixteenth and seventeenth centuries, which gave birth to the modern scientific enterprise.[7]

Beginning in 1970, the amount of science reported in the media exploded. In the United States, the 1970s and 1980s saw the creation of science sections in dozens of newspapers across the country, the launch of multiple glossy popular science magazines, and the inauguration of a new weekly television series—*Nova*—devoted to science. Popular science books reached a significant point in the mid-1970s. Before then, there were rarely more than ten titles in the *New York Times* best seller list each year. But afterward, there were rarely fewer than ten best sellers each year. The situation in Britain was similar. Science flowed through popular culture.[8]

Television allowed scientists to speak to vast numbers of citizens. The BBC series *The Ascent of Man* told a science-based story of human history. Broadcast in Britain and the United States in the early 1970s, it was hosted by mathematician and intellectual Jacob Bronowski, who had written and spoken about science to wide audiences in magazines and television long before the show granted him international prominence. During the same decade, across the Atlantic, a planetary scientist was proving himself an engaging media presence, a scientist who would became his era's best-known public scientist: Carl Sagan.

Sagan symbolized an era when the television age met the space age. He was a planetary scientist at a time when space became a proxy battleground for rival Cold War superpowers. He was telegenic at a point where it was clear that television favored personalities, like him, who were articulate, attractive, eloquent, and enthusiastic. He was already well known at the end of the 1970s as a Pulitzer Prize–winning popular science writer who regularly explained astronomy to the hundreds of thousands of nightly viewers of *The Tonight Show* with Johnny Carson.

But when he unveiled the universe to half a billion viewers in the 1980 television series *Cosmos*, he was propelled to unprecedented global fame. Viewers in sixty nations followed the planetary scientist on his thirteen-part personal odyssey through eons of cosmological and human history. His spin-off book of the series, *Cosmos*, spent more than seventy weeks on the *New York Times* best seller list and earned him more than $1 million in royalties. *Time* featured Sagan on its cover and called him a "Showman of Science," "the prince of popularizers," "the nation's scientific mentor to the masses," and "America's most effective salesman of science."[9]

A producer of *Cosmos*, Adrian Malone, vowed to "make Carl a star." And indeed the show led to a surge in media and public attention paid to Sagan. Journalists reported on his personal life, writing about his trademark turtlenecks and his distinctive orange Porsche 914 with its license plate, PHOBOS, one of the moons of Mars. He had to cope with the women who appeared at studios demanding to see him, convinced he spoke directly to them through their television screens. He sometimes sat facing the wall in restaurants to avoid the stream of autograph hunters and well-wishers.

His celebrity brought lucrative rewards. The $2 million he received for *Contact*, his 1985 novel about the scientific search for extraterrestrial life, was, at the time, the largest advance ever given by a publisher for a work not yet in manuscript form. It also brought him influence, granting him a public platform for his anti-nuclear advocacy, as he warned political leaders about the devastation that would occur in the radiation-soaked darkness of a global nuclear winter.[10] Students who watched *Cosmos* wanted to become scientists. No modern scientist had yet achieved such reach, renown, and reputation.

But his fame damaged Sagan's standing in the scientific world. Harvard denied his bid for tenure, a lifetime appointment that a university awards to accomplished scholars. The nation's most prestigious scientific society, the National Academy of Sciences, rejected him as a member. A number of influential peers dismissed him as a mere popularizer and not a real scientist, someone who spent too much time on *The Tonight Show* and too little time engaged in the painstaking grind of observing the planets.

He came to starkly illustrate a feature of modern scientific fame, a feature that critics later called the "Sagan Effect": the perception among researchers that the level of scientists' public fame was in direct opposition to the quality of their research work. Popular scientists, in effect, were not seen as strong scientists. Before his media career, however, Sagan had established a sound reputation as a researcher, known for his pathbreaking work that explained how Venus became boiling hot and violent windstorms raged across the surface of Mars. He accumulated five hundred career publications—an astonishing rate of productivity that averaged one published academic paper each month. The Sagan Effect, for Sagan, was false.[11]

Not that Sagan was the only scientist to spot the media's enhanced power. He was one of several scientists in U.S. public life in the 1960s and 1970s who saw the media as a way to influence public and political attitudes to science. These "visible scientists"—including anthropologist Margaret Mead, biologist Paul Ehrlich, and chemist Linus Pauling—broke with conventional ways to shape science policy. They bypassed the traditional ways that experts gave behind-the-scenes advice to policymakers. They went directly to the public instead, using the mass media to put science on the public agenda and therefore shape citizen attitudes and, as a result, affect science policy. They showed that the individual scientist working in a cutting-edge area of science, once they were sufficiently articulate, controversial, and distinctive, could attract and hold the media spotlight. [12]

These visible scientists ruptured the conventional ways researchers earned scientific and public attention. As described by a founding father of the sociology of science, Robert K. Merton, an individual scientist's reputation was traditionally established exclusively *within* science. A scientist gained recognition only after their published research was validated by their peers. The more and better their research, the more their reputation grew, the greater their status in science. The ultimate accolade was the Nobel Prize, the public symbol of scientific excellence, a public award bestowed on those researchers deemed to have produced the world's best science. [13] But Sagan and other visible scientists had a reputation that was in part created *outside* science. As well as scientific credentials, what also mattered was how they communicated, how engaging they were, how their science was tied to public issues, and how interesting they were as personalities.

A NEW CELEBRITY CULTURE

As media personalities, the visible scientists were early actors in what has become a pervasive celebrity culture. Today the media concentrate on personalities, representing complex events and issues through the prism of personalities. As cultural historian Leo Braudy noted in his history of fame, *The Frenzy of Renown*, in our celebrity culture "human faces are plastered on every idea and event" and "complex phenomena wear the reduced features of emblematic individuals." [14] Critics and commentators have lamented this shift in culture, viewing the media's obsession with celebrity as the triumph of the trivial, the elevation of the inane, the proof of a debased and dumbed-down culture.

But there is a more positive view of fame, one that sees great power and importance that celebrities hold in our media-saturated public life. For Braudy, fame "sits at the crossroads of the familiar and the unprecedented, where personal psychology, social context, and historical tradition meet." [15] Stars were once seen as a powerless elite: they had recognition but no influence. [16] That idea has long passed. Now celebrities have power because they vividly represent ideas, issues, and ideologies, allowing people to visualize and make sense of abstract concepts. As the author David Foster Wallace wrote about sports stars, "Great athletes are profundity in motion. They enable abstractions like *power* and *grace* and *control* to become not only incarnate but televisable." [17]

Moreover, celebrities personify and act as figureheads for social movements. Celebrities with enduring popularity and prominence have a way of portraying the deep questions, tensions, and conflicts of their eras. Celebrities come to personify the culture and society of their particular time and place. They help people make sense of the world.[18]

Scholars of communication and popular culture, however, have defined celebrity in a technical sense. Celebrity is a phenomenon formed as a result of three interconnected processes. The first process can be seen when someone is portrayed in media coverage as a distinctive individual whose public and private lives merge. (When Sagan died, the *Australian* said he had had the good luck to have "a compelling presence" and "good looks."[19])

The second process is when a person becomes a cultural commodity, used to sell his or her own work, but also potentially as a way to advertise other cultural products. (The *New York Times* said *Cosmos* focused so intensely on its charismatic host that the show could have been subtitled: "The Selling of Carl Sagan."[20]).

The third—and most complex, but arguably central—process concerns the way the person comes to represent and embody ideas, ideologies, and processes. Sagan embodied for many the idea of the scientist, the heroic seeker after truth who sought to overthrow ignorance and superstition with rationality, showing how science was, in his words, a candle in a demon-haunted world.[21] (When he died, the *Atlanta Journal and Constitution* said: "For the common man, his was the face of science."[22]) Describing this last process, scholar and writer for the *New Yorker* Louis Menand wrote that the height of a celebrity's stardom comes when their personality intersects with history in "a perfect congruence of the way the world happens to be and the way the star is."[23]

The process through which a public figure becomes a celebrity—how media attention focuses on their private life, how they become commodities, how they symbolize wider cultural issues—has been labeled with a useful, but somewhat cumbersome, neologism: celebrification.[24]

NEW VIEWS OF SCIENCE

The long sixties also brought into the public spotlight the misuse of science. Rachel Carson's *Silent Spring*, first published in 1962, vividly demonstrated that science can damage society and public life. A zoologist who worked with what became the U.S. Fish and Wildlife Service, Carson argued that the annual uncontrolled release into the wild of five hundred potent chemicals, many of them insecticides, poisoned the landscape, destroyed delicate ecosystems, and killed species.[25] *Silent Spring* galvanized the nascent environmental movement. It prompted President John F. Kennedy to order an investigation into the claims. The book also inspired many science journalists to become more aggressive and escape their dominant historical role as disseminators of stories that told only of science's triumphs and virtues, a stance that one reporter said made them "science propagandists."[26] *Silent Spring* had another effect. Before the book, when experts disagreed about scientific problems, their discussions usually occurred in closed conference rooms or within the hermetically sealed confines of academic journals. But *Silent Spring* brought these debates into the pages of the press and the unforgiving glare of live studio debate.

The long sixties in the United States, particularly, exhibited a bifurcated attitude toward science. After the Soviet Union launched the satellite Sputnik 1 into orbit in 1957, the quest to dominate space became a national political obsession—even if, contrary to some modern myths, public opinion surveys showed that the event did not inspire a surge in public understanding of science. The Apollo mission to the moon demonstrated the awe-inspiring prowess of science and technology. Yet on Earth, the counterculture movement viewed science as being co-opted by industry for nefarious ends, including nuclear weapons, and napalm used in the Vietnam War.

In the same period, historians, sociologists, and anthropologists peered into the hidden internal workings of science. What they found shattered "The Legend," the modern myth that science was an orderly process where bricks of new insight built on blocks of established knowledge to construct a progressively taller temple of truth. According to The Legend, scientists were part of a great "story of a protracted struggle, which will never end, against the inertia of superstition and ignorance, against the liars and hypocrites, and the deceivers and the self-deceived, against all the forces of darkness and nonsense."[27]

But historians and sociologists attacked this myth. *Silent Spring* was published the same year as *The Structure of Scientific Revolutions* by physicist and historian Thomas Kuhn, who said science was not progressive or cumulative. Science instead advanced through revolutions where old sets of ideas—paradigms—were replaced in a process called a paradigm shift by a new set of understandings about how science should be undertaken.[28] Other researchers took the skills of anthropologists who recorded the lives of indigenous peoples and ventured into laboratories to document the reality of how scientists did their day-to-day work. The studies showed scientists were methodological, professional, and socially conscious. But at the same time, they were shown to be ambitious, greedy, and status-obsessed. They were jealous of their rivals and sought the approval of their peers. Scientists, in short, were like everyone else.

In the long sixties, the scientific enterprise changed fundamentally. Research became increasingly privatized, entrepreneurial, and profit focused. Researchers produced knowledge not only in universities, but also in institutes and industrial laboratories—organizations that today collaborate within and between nations. Science became increasingly fragmented, splintered into more and more fields and subfields, turning scientific experts into ultra-specialized researchers who turned out vast amounts of knowledge. About 1.8 million scholarly articles were published in 2012 in approximately 28,100 peer-reviewed journals. Publicity mattered, too—as it brought favorable attention to science and showcased its achievements to society. Individual scientists realized they must sell their science to funders, journalists, and the public that often pays for their work.

Exemplifying this new science in the 1970s was Nobel Prize–winning geneticist James Watson. In *The Double Helix*, his 1968 autobiographical account of how he helped determine the structure of DNA, Watson painted himself as an individualist, hungry for success, sex, and celebrity. Watson was a new kind of scientist, a reviewer of the autobiography lamented in *Science*, one "subject to, and forming part of, all the vulgarities of the communications media."[29]

A CRISIS OF TRUST IN SCIENCE

As the twentieth century ended, the scientific community felt itself besieged.

The entire idea that science could determine truth was ridiculed by writers who formed part of the era's defining intellectual trend—postmodernism. The wide-ranging term described a world where the nature of reality depended on individual interpretation, where what counted as knowledge constantly changed and where traditional ways of coming to a unified understanding of the world—philosophical ideas, religious ideas, scientific truths—were no longer considered valid. No one truth explains everything.[30]

Carl Sagan argued in his 1995 book *The Demon-Haunted World* that science was under threat from a mass of irrational forces such as superstition, pseudoscience, and advertising. The same year, historian of science Gerald Holton, in a book subtitled *The Rebellion against Science at the End of the Twentieth Century*, argued the era was characterized by a rebellion by politicians, academics, and writers against the idea that science can lead to progressively sounder, objective knowledge that is based on rational thought and can improve wider society. For Holton, this constituted no less than an all-out assault on the foundations of Western civilization. In peril were the legitimacy and authority of science, as well as the historical ideals of reason, rationality, and truth.[31]

But these somewhat alarmist accounts played down the major controversies that undermined public faith in science. A salmonella outbreak in Britain dented trust in the food industry. That trust was broken with the outbreak of mad cow disease, a condition that scientists and politicians first told the public could not infect humans, a distrust that spilled over into vociferous objections to genetically modified food. In the United States, the near meltdown at Three Mile Island in 1979 highlighted the dangers of nuclear power. A reinvigorated creationism emerged in the 1980s and morphed into intelligent design in the 1990s. During the two terms of President George W. Bush, liberals said his administration mounted an apparent war on science, especially in climate science and stem cell research. Vaccination rates fell after public panic greeted the publication of a now-discredited scientific paper that found a link between the MMR vaccine and autism.

No wonder the public in the Western world remained confused or skeptical of science. Official survey figures from 2007 showed only one in four Americans grasped the basic concepts that would allow them to read the science section in the *New York Times*. Four out of five adults knew that the center of Earth is very hot. Only 63 percent of adults knew that Earth goes around the Sun once each year. A majority of American adults did not believe humans share a majority of genes with chimpanzees or mice. Only 30 percent understood or accepted the theory of the big bang. Only 40 percent accepted evolution, a number that has declined over the last two decades. Half of adults believed that humans existed at the same time as dinosaurs.[32]

Citizens at the end of the last century had different attitudes to science. They were optimistic, pessimistic, conflicted, or disengaged.[33] They were uneasy about how developments in areas like biotechnology raced ahead of public awareness and public consent, areas where knowledge is not as certain as knowing the Earth goes around the sun or the center of the planet is indeed very hot.

Policymakers were alarmed, describing the end of the century as a crucial period for science's status in society. A report from the House of

Lords—the upper house of the UK Parliament—described a crisis of confidence and a crisis of trust in science. It wrote in a 2000 report that "public unease, mistrust and occasional outright hostility are breeding a climate of deep anxiety among scientists themselves."[34] Without public understanding of science, feared scientists, citizens could not take part in democratic life. No understanding meant no public support for scientists. No understanding meant no legitimacy for science policy. In response, scientific elites understood that they had to be more proactive in the public arena. Capturing this idea, two science communication scholars wrote in a classic 1998 text *Science in Public*: "In the last decade or so, scientists have been delivered a new commandment from on high: *thou shalt communicate*."[35]

The scientific community correctly viewed the media as the key to potentially enhancing the public's scientific literacy. But their understanding of how communication worked was often simplistic. Their argument ran: scientists get the science right, journalists and other media professionals communicate it in simple yet accurate ways, and the public comes to not just understand science, but also to appreciate it. In this view, scientific literacy, or the public understanding of science, has traditionally been viewed by scientific elites as knowing more of the scientific facts that citizens should have learned at school. The ultimate aim is to respect scientific authority, support research funding, and come to see the world and its problems in the same way as scientists do.

That simplistic view fails to capture the broader dimensions of scientific literacy. It is not just memorizing facts like citizens did at school. It is understanding science as a method of understanding the world, a body of knowledge, a source of technology, a creator of ways of thinking, a means of challenging established ideas. Science literacy, scholar John Durant has argued persuasively, consists of three connected levels. The first and most basic level is indeed *knowing a lot of science*: understanding fundamental facts, concepts, and theories. The second deeper level of scientific literacy means *knowing how science works*: knowing the process of scientific inquiry, such as the different methods scientists use, the importance of evidence, the operation of peer review. The third and deepest level of literacy means *knowing how science really works*: understanding how science is undertaken by individual scientists subject to the same drives and pressures as other workers.[36]

Related to scientific literacy is another misunderstood term: science popularization. The simplistic view of popular science books, for example, sees them only as media that translated specialist science to nonscience citizens. Popular science books do, of course, explain complex ideas. But they also push scientific advance. *Chaos: Making a New Science*, published in 1987, helped scientists in various fields understand the emergent ideas of chaos theory that used advanced mathematics and physics to explain complex systems. Books like *Contact* or *The Double Helix* recruit students into science. Books like *Silent Spring* act as public forums where science-related issues can be aired and discussed.[37]

The scientific community became publicity savvy in other ways. Scientists have acted as consultants to Hollywood films, providing the entertainment industry with knowledge about science that can enhance their on-screen stories. Public relations professionals work in scientific institutions, carefully crafting messages to make their brand of scientific work look good in public. Journals like *Nature* seek to not only publish important research, but also to have that research talked about by other media. Science, consequently, has been dragged into the media-saturated public arena.[38]

THE NEW CELEBRITY SCIENTISTS

The scientists who made science matter for the public were public intellectuals. They wrote for audiences beyond their professional communities. They spoke in public forums. A scientist becomes a public intellectual by following a four-step process. First, he or she becomes an expert in their specialist subject. Then the scientist accesses media channels that allow them to speak to general audiences outside their sphere of expertise. Third, he or she expresses views or opinions on topics and themes that engage or otherwise articulate the concerns of those broad audiences. Finally, the would-be intellectual establishes a reputation for voicing those views and opinions in an interesting and important fashion, and not being shy about doing so through broad-access media . . . and a public intellectual is born.[39] Public intellectuals do not work solely within a *professional culture* of other credentialed experts. They also work within a broader *public culture* that includes experts from other fields, journalists, writers, critics, and citizens.[40]

Scientific public intellectuals, like other public intellectuals, have a powerful cultural role. The eminent cultural critic Edward Said, in *Representations of the Intellectual*, wrote, "The intellectual is an individual endowed with a faculty for representing, embodying, articulating a message, a view, an attitude, philosophy or opinion to, as well as for, a public."[41] The intellectual also exists as a "figurehead or spokesperson or symbol of a cause, movement, or position."[42] The intellectual testifies on the public stage, articulates a cause or idea to society, and is "someone who visibly represents a standpoint of some kind."[43]

At the end of the twentieth century, the scientific public intellectual became a fixture on the cultural landscape. One book argued that they overcame the binary opposition identified by scientist-turned-writer C. P. Snow between the science and humanities—the two cultures—and instead scientific intellectuals formed a new third culture. But considering the unsurpassed influence science has had on modern life—power stations, processed food, penicillin, painkillers—a more accurate description of the place of science in society was provided by British journalist and author Bryan Appleyard when he wrote: "I think there is one culture dominated by science and defined by different attitudes to science."[44]

In this atmosphere, a small handful of scientists came to dominate public discussion of science, publishing bestselling titles, receiving six-figure advances for books about esoteric topics like quantum physics, producing science documentaries, contributing to late-night talk shows, appearing in glossy magazines, being photographed by celebrity photographers, lobbying Congress. The *Independent* newspaper in the United Kingdom encapsulated the new trend when it said that the turn of the twenty-first century saw science dominated by its "media superstars." *Vogue* magazine in 1997 said serious science had become glamorous. *Current Biology* said scientists were portrayed as "stylish and even sexy."[45] Scientists became public intellectuals and, in the process, became celebrities.

In the chapters that follow, I chronicle the public lives and endeavors of eight scientists who evolved—in the fierce glare of modern media—into major scientific celebrities. I then combine their individual stories to tell a larger tale: the power of the new scientific celebrity to sell ideas, shape public understanding of science, catalyze social movements, influence policy, and change the course of scientific research.

The public lives of the compelling figures I analyze also show the sometimes strange ways scientific ideas spread through culture, how particular scientific ideas come to resonate deeply with the public, and how certain figures come, over time, to be deeply connected with citizens, fulfilling a deep and not always clearly articulated cultural need. By demonstrating the complex dynamics of how and why a particular scientist became a celebrity, I offer an answer to an intriguing question posed by one *Slate* science writer who pondered the public profiles of two physicists: "what does it say about the cross-talk between science and popular culture that Brian Greene is a celebrity and Freeman Dyson is not?"[46]

The eight celebrity scientists I profile in this book, including Greene, are:

- **Stephen Hawking.** The cosmologist became the world's best-known contemporary scientific celebrity after the success of *A Brief History of Time*, first published in 1988, brought cosmology to millions of readers worldwide, but his subsequent public career has been marked by a succession of private revelations, recycled versions of his best selling popular book, and often caustic public evaluations of his scientific reputation.
- **Richard Dawkins.** Called "Mr. Public Science himself," "Professor Evolution," and "Professor Science," the evolutionist and writer fashioned his fame over his decades-long public career as a strident advocate of evolution, combative defender of science, and relentless critic of religion.
- **Steven Pinker.** Once described as "famously rock 'n' roll with his long, curly hair and his cowboy boots," the Harvard cognitive scientist explained the biological roots of language, and argued controversially that biology plays a major role in molding not just human behavior, but human society and culture.
- **Stephen Jay Gould.** The late paleontologist, described by one critic as a "learned Harvard professor and baseball-loving everyman," enthused millions about evolution, battled creationism, and tried to reconcile science and religion.
- **Susan Greenfield.** The former Oxford professor of pharmacology and member of the UK House of Lords demonstrates a contentious portrayal of female scientists—she has been called a "mini-skirted media celebrity"—and an ability to raise and discuss uncertain science-based social problems, such as the claimed harmful effects that screen technologies have on children.
- **James Lovelock.** Called the "intellectual guru of the environmental movement," the independent scientist worked for decades in two isolated English farmhouses and showed with his controversial Gaia theory of the living Earth that a popular science book can not only influence science, but also spark an entire belief system, and come to powerfully symbolize the current climate crisis.

- **Brian Greene.** The physicist is the public face of string theory, the novel branch of physics that captured the turn-of-the-century scientific and public imaginations, and the scientist who can move seamlessly between speaking at academic conferences and appearing on *The Big Bang Theory* without losing his scientific status.
- **Neil deGrasse Tyson.** The director of the Hayden Planetarium in New York became the contemporary heir to Carl Sagan, and is the unofficial chief public spokesman for science and space science in the United States, shaping public attitudes, science policy, and the future of his field.

I deliberately chose these eight scientists as subjects for case studies. I anticipated they would be telling and revealing examples of scientific celebrity and would demonstrate clearly the impact of fame on culture and society. All the scientists I chose have much in common. Each one made a deliberate decision to enter public life. Each went on to maintain a regular presence on the public agenda, writing popular books, participating in television broadcasts, and being the focus of voluminous journalistic writing and frequent scholarly analysis.

Yet the scientists are different in important ways. They are specialists in a variety of fields and subfields. They have worked in different roles in various types of institutions. As examples, Hawking is a research cosmologist at an elite university, while Tyson leads a major public planetarium, and Lovelock is an outlier in modern science in that he spent the majority of his career as an independent scientist. Additionally, one of the selected scientists is a woman and another is African American, reflecting the paltry but important diversity present in star scientists. All have operated largely in Britain and North America, the areas that are most familiar to me and so allow me to delve deep into their particular cultures.

To be sure, I could have profiled other scientists—such as physicist Richard Feynman or primatologist Jane Goodall—instead of, or in addition to, the eight selected. These and other scientists with various degrees of public prominence make guest appearances in the stories that follow. But I judged that their routes to fame and their impact on public life, while in many ways unique and particular, would follow similar paths to those profiled here. There is one exception: the rapid rise of the most recent celebrity scientist in the United Kingdom—physicist Brian Cox—is addressed in the concluding chapter, as he represents in some ways a new iteration of the celebrity scientist.

There are other studies, including biographies and autobiographies, that have examined the lives and works of my chosen star scientists. I have woven their insights into my critical profiles, as these accounts also contribute to the stars' public images. No studies, however, examine the evolution of their stardom in the way I do in the following chapters. Now, anyone writing a book about science and scientists must make clear where they stand in relation to science. Stephen Jay Gould best articulated my position when he described his view of science in his 1996 book *Full House*. "Nature is objective, and nature is knowable," he wrote, "but we can only view her through a glass darkly—and many clouds upon our vision are of our own making: social and cultural biases, psychological preferences, and mental limitations (in universal modes of thought, not just individualized stupidity)."[47] For me, this is a view that best shows how science really works.

Each of the following chapters presents a chronological profile of one scientist. I examine and explain their rise to fame using as tools the concepts of celebrity and the public intellectual. I trace their trajectories as public intellectuals by examining their books, broadcasts, journalism, and other activities in public life, such as speeches and talks. I position their works within a wider stream of history and place them within their historical context. I examine their development as celebrities by seeing how their public image is crafted from four core types of media: how they are portrayed in their books and other writings and broadcasts, how they are portrayed in publicity material such as interviews and profiles, how they are represented in promotional material such as blurbs and press releases, and how they are described in critical and scholarly writing. For some chapters, I interviewed the scientists and those who helped construct their public image. I ultimately aim to provide rich, detailed, and readable profiles of some of the iconic figures of modern science.[48] Unlike new approaches that are being developed to examine the scientists over the past two centuries to create a grand Science Hall of Fame, I keep my focus on a close analysis of prominent media scientists of the last several decades.[49]

The order of the chapters reflects the different ways that scientists exploited and managed their fame in their careers. Dawkins used his fame to move far from the laboratory and become the head of a new social movement of atheists. Pinker and Gould managed the difficult task of being famous public intellectuals and prolific university-based researchers. Greenfield embraced fame and its advantages for scientific reputations. Lovelock reluctantly embraced stardom after he was shut out of mainstream science.

Greene uses his fame as a passport to move seamlessly between the worlds of science and entertainment, while Tyson shows the power of a public scientist, one who has not got a strong record of scientific research, to influence public understanding, scientific debate, and public policy.

But I begin by focusing on a scientist who worked during the 1960s on the exciting and emerging science of black holes, the scientist whose book overtook *Cosmos* as the bestselling title ever, the scientist who shows us most vividly the features of scientific fame: Stephen Hawking.

TWO

The Paradoxical Fame of Stephen Hawking

On January 4, 2012, four days before Stephen Hawking turned seventy, *New Scientist* published an exclusive interview with the man it called "one of the world's greatest physicists." Asked about the most exciting development in physics during his lifetime, Hawking said it was the confirmation of the big bang. Asked to name his biggest scientific blunder, he said it was his mistaken view that black holes destroyed the information they swallowed. Asked what he thought about most during the day, he said: "Women. They are a complete mystery."

News outlets worldwide—including CBS News, the *Guardian*, the *Telegraph*, the *Huffington Post*, and the *Hindu*—angled their reports about Hawking's birthday around the sound bite. The quip sparked a week of Hawking coverage that culminated in a special symposium organized by Cambridge University to celebrate the life and work of its best-known academic, marking a historic day that few believed Hawking would live to attend.

Doctors predicted he had not long to live after he was first diagnosed as a twenty-one-year-old undergraduate with amyotrophic lateral sclerosis (ALS), or Lou Gehrig's disease, a form of motor neuron disease that shreds the nerves controlling the body's muscles—leading to their progressive wasting and weakening.

Yet he reached his seventieth year and more than one hundred prominent physicists attended the exclusive event, alongside model, actress, and
Cambridge graduate Lily Cole and entrepreneur Richard Branson, who told
newspapers that Hawking should "have won the Nobel Prize many times"
and "is someone who has discovered many things in his lifetime."

Hawking himself did not attend due to illness, but a prepared speech—
titled "A Brief History of Mine," a title recycled from other public speeches he
had delivered—was carried through the auditorium and around the world by
journalists for the news wires Agence France-Presse and Reuters. Even without his presence, his birthday was an orchestrated media spectacle, a demon-
stration of how completely he was enmeshed within the culture of celebrity.[1]

Hawking—"the most famous scientist of our time," "a celebrity, a part
of popular culture," "the Mick Jagger of theoretical physicists"—is without
doubt the most famous scientist of the modern era. His rise to pop suprem-
acy demonstrates most vividly the characteristics and causes and the possibilities and pitfalls of scientific celebrity. It is a rise that began in Cambridge,
as scientists in the sixties began to gaze with renewed interest at the cosmos.[2]

THE RISING STAR OF
RELATIVITY'S RENAISSANCE

Until the 1960s, most physicists viewed cosmology as a peripheral scientific
field—if indeed they considered it scientific at all. The scientific study of the
universe at large occupied a midpoint between science and philosophy. It
had metaphysical dimensions and its theories could not be tested by observation. But it had a rich scientific heritage. Its foundational concepts were
based on Einstein's general theory of relativity: his theory of gravity that
described the motions of galaxies. Yet few researchers worked to develop
the general theory, because the mathematics involved were too difficult and
its ideas could not be tested in experiments. It was difficult to find strong
evidence to back up new concepts.

As a consequence, in the late 1950s, cosmology's base of evidence
was so uncertain that physicists could not discriminate between two rival
theories about the formation of the universe. One theory proposed the evolutionary view of the universe: the cosmos exploded into life ten billion years
ago and has expanded after a moment of creation that became known as the
big bang. The rival explanation—the steady-state theory—proposed a static

universe with no beginning and no end. Matter was continually created. At the time, different observations favored one or other theory. No evidence proved sufficiently persuasive to settle the debate.

But in the 1960s, new observational methods and instruments turned the universe into a laboratory. The observations found the universe was filled with cosmic background radiation, the physical residue of the universe's violent beginning predicted by the big bang theory. The new evidence proved steady-state theory wrong. The observations also found several phenomena—such as quasars, star-like objects brighter than entire galaxies—that could only be explained by Einstein's general relativity. The theory had a newfound usefulness in that it could explain fundamental features of the universe. With these advances, cosmology came of age. The number of scientific articles published in the field between 1962 and 1972 increased from about 50 to 250. When Hawking graduated with his PhD from Cambridge in 1966, he was at the forefront of this renaissance.[3]

Hawking dove into solving a fundamental problem: What happened at the origin of the universe? To tackle this question, he focused on one interesting and mysterious cosmological phenomenon—the black hole. This was the condensed remnant of a dead star, with a gravitational field so powerful that not even light could escape. The center of a black hole becomes a cosmic dead-end that physicists call a singularity, an uncharted territory where the predictable laws of physics break down.

But cosmologists faced a major problem. They worked with general relativity, which describes the large-scale structure of the universe. But it could not explain what happened in a singularity. For that, it needed to merge with quantum theory, which described the small-scale workings of the universe. Needed was a unified explanation. Needed was a unified theory of quantum gravity. Needed was a theory of everything.[4] In the 1970s, Hawking sought this synthesis. In the process, he made his signature contribution to cosmology. In 1974, he published a landmark paper in *Nature* that demonstrated black holes did not suck up everything—they emitted a form of heat that would eventually be called Hawking radiation. For this significant finding, he was admitted to the Royal Society, Britain's most prestigious scientific society. At age thirty-two, Hawking's professional status was already sealed.

Meanwhile, black holes resonated in 1970s culture. For *Time*, they were a crossroads between astrophysics and metaphysics, a point where "science finally converged with religion," part of a cultural landscape filled with a "faddish craze for the likes of parapsychology, the occult, UFOs, thinking plants . . . and other pseudoscientific hokum."[5] Popular science writers mixed these scientific, spiritual, and religious themes in a slew of popular cosmology books published during the 1970s and 1980s, such as *The Omega Point, God and the New Physics, The Dancing Wu Li Masters*, and *The Tao of Physics*.

As an expert in black holes, Hawking first appeared as a source in specialist science media.[6] *Science News* in 1973 called Hawking "one of the foremost experts in the field." Notably, the report did not mention his physical appearance.[7] Hawking was one of several cosmologists interviewed in the 1975 BBC documentary *The Key to the Universe*. A 1977 article he wrote for *Scientific American* not only popularized Hawking radiation,[8] but also positioned his own black hole research as part of the quest to find a theory of everything.[9] It was not long before more mainstream media saw Hawking as more than just an expert source.

PROFILING THE EMBLEMATIC PHYSICIST: THE DISEMBODIED BRAIN

Magazines found Hawking irresistible. *New Scientist, Time, Omni, Science 81, Reader's Digest*, the *New York Times*, and *Vanity Fair* each profiled him in-depth between 1978 and 1984. These portrayals were crucial for the establishment of Hawking's image because magazines, like no other media, capture and reflect social trends. He was the perfect symbol for the strange, otherworldly new physics. Their accounts crystallized a particular image of Hawking, with his muscular mind in a broken body, as the new cosmology's emblematic physicist.[10]

Writers portrayed Hawking as separate from his body—as a disembodied brain. In an article headlined "The Unfettered Mind," *Science 81* told its 50,000 readers: "By the early 1970s, Hawking was confined to a wheelchair. But his mind was soaring."[11] *New Scientist* wrote that "much of the most outstanding work on black holes has been performed in the head of a remarkable Cambridge physicist. . . . The work *has* to be done in his head, because Hawking is crippled by a disease that confines him to a wheelchair and renders him unable to write."[12] Over time, Hawking, according to the *New York Times*, was becoming increasingly "a cerebral being."[13] In an edition that featured articles on singer Boy George, writer Gore Vidal, and academic Umberto Eco, *Vanity Fair* introduced Hawking as breaking free of his body: "Unable to write, or even to speak clearly, he is leaping beyond relativity, beyond quantum mechanics, beyond the big bang, to the 'dance of geometry' that created the universe."[14]

Profiles also found Hawking the perfect way to visualize cosmology's mystical dimensions. The science and science fiction magazine *Omni* called him "The Wizard of Space and Time" and told its 750,000 readers that his black hole research has "shaken the world with a discovery so weird and still so mysterious that even a statement of one of his findings sounds like a Zen koan: 'When is a black hole not black? / When it explodes.'"[15]

But even as writers described him as a disembodied mind, they examined his domestic life. *New Scientist* observed that the bond between Hawking and his ten-year-old son, Robert, "is particularly strong. . . . The two spend hours working out problems on Robert's programmable pocket calculator."[16] *Time* described him as a "devoted" father and noted that his wife "to whom he has been married for thirteen years, often accompanies him to scientific meetings."[17] *Science 81* published a photograph of Hawking and his sons, Tim and Robert, playing with "a model of a black hole."[18] *Omni* described how Hawking relied on help to eat, dress, write, comb his hair, and fix his glasses.[19] *Science 81* noted that his office was like any other physicist's office: "filled with physics texts, papers on cosmology, a blackboard with scribbled equations, a computer terminal, a tidy paper-laden desk, and pictures of three handsome children."[20]

One science writer warned about the consequences of Hawking's rising profile. Timothy Ferris feared Hawking's image could become detached from his scientific work. "Hawking's story promises—or threatens—to make him a celebrity," wrote Ferris in his *Vanity Fair* profile. "By offering up a Hawking sans physics, the press threatens to turn him into an adult version of a muscular-dystrophy poster child, an afflicted soul who smiles sweetly and demonstrates great courage without being expected to have anything much to say."[21] Ferris reviewed the first book on Hawking, *Stephen Hawking's Universe: An Introduction to the Most Remarkable Scientist of Our Time*, and caustically observed that the book's author said Hawking was "upon the basis of what research I cannot imagine . . . 'our planet's most fully developed cerebral creature.'"[22]

The image of the scientist as a cerebral creature has long historical precedent. Isaac Newton's dominant public image, for example, was that of "a disembodied mind in communion with natural or divine truth."[23] It conveyed a particular view of scientific thought. Scientific research was not something carried out by workers in laboratories, wrestling facts from nature. Instead it was solely cerebral and otherworldly, an activity that aimed to grasp transcendent truth only by pure thought.[24] Some figures were able to escape their bodies to grasp this otherworldly truth: disembodied minds.

As Hawking soaked up media attention, his scientific career advanced. In 1979, he became the Lucasian Chair of Mathematics at Cambridge, a position once held by Isaac Newton. His inaugural lecture, "Is the End in Sight for Theoretical Physics?," predicted that it was a realistic possibility that by the end of the twentieth century, physicists would establish a unified theory that would encompass all theories in the field—essentially creating a theory of everything that would connect the ideas of general relativity to quantum mechanics to explain the workings of the universe. If that happened, there would be no more fundamental problems to solve. It would mean the end of physics.[25] The theme of the lecture would also be the big idea for a popular book he planned to write.

THE SELLING OF HAWKING

Hawking wanted *A Brief History of Time* to make money. A best selling general-interest guide to the cosmos, he hoped, would help pay his daughter's school fees and the escalating cost of his health care. Wanting the book sold in "airport book stalls,"[26] he rejected an offer from Cambridge University Press, which offered an advance of £10,000—until then the largest it had negotiated with an author. The agent sold it in a telephone auction to trade publisher Bantam Books, which offered a $250,000 advance and generous royalties. When he first met Bantam Books editor Peter Guzzardi in 1985, Hawking's opening words were: "Where's the contract?"[27]

Bantam offered the commercial focus Hawking wanted. Hawking was just the type of salable author Bantam desired. It became one of the world's major publishing firms in part because of its aggressive marketing and attractive cover art. Its commercial growth in the 1980s and 1990s was driven by a strategy centered on the hardback publication of "blockbuster nonfiction."[28] Bantam convinced Hawking the title was right after he initially found it flippant.[29] An editor advised Hawking to purge an initial draft of scientific formulae because, he said, each equation would "halve the sales."[30] The same reasoning prevailed when Hawking considered deleting the book's controversial last line saying that a unified theory of science would be "the ultimate triumph of human reason—for then we would know the mind of God"[31]: "Had I done so, the sales might have been halved."[32]

The publishers promoted Hawking around ideas of disembodiment. "From the vantage point of the wheelchair where he has spent the last twenty years trapped by Lou Gehrig's disease," read the dust-jacket description of the U.S. edition, "Professor Hawking himself has transformed our view of the universe." Describing the author, the UK edition said: "Confined to a wheelchair for the last twenty years by a motor-neuron disease, Professor Hawking is best known for his work on black holes." The same edition featured on its cover a photograph of a black-suited Hawking sitting in his wheelchair, superimposed before a starry universe.[33]

A Brief History of Time did not seem like a sure-fire best seller. A 198-page cosmological guide for general readers, the book explained a series of concepts in high-end physics, such as the big bang, general relativity, quantum physics, and black holes, and it ended with short biographical sketches of Einstein, Galileo, and Newton. After it was first published in 1988, booksellers could not keep enough copies in stores. It stayed on the best seller lists of the *New York Times* for 147 weeks and on the *Sunday Times* for 237 weeks. In just over three years since its publication, it sold more than four and a half million copies. It would go on to sell more than ten million copies and be translated into forty languages as it overtook *Cosmos* as the most popular science book ever.[34]

Book critics, like magazine editors, found Hawking fascinating. Martin Gardner in the *New York Review of Books* said: "Before discussing his stimulating book . . . I shall say something about the book's even more extraordinary author."[35] Robyn Williams in the *Sydney Morning Herald*'s review wrote in the first paragraph: "I'm going to resist the almost unbearable temptation to tell you first about the author. He is certainly unique, tragic and triumphant."[36] Although this book was clearly not intended to be an autobiography, said the *New York Times*, "it is still disappointing that Mr. Hawking keeps such revelations to a minimum."[37] Hawking later dismissed most of the reviews; they almost invariably followed the same formula of saying that he suffered from motor neuron disease, was in a wheelchair, could not speak, could only move his fingers, yet has written a popular book on one of the fundamental mysteries of the universe: its formation. The human interest story helped the book be popular, he said, "but it was intended as a history of the universe, not of me."[38]

The book's success puzzled commentators. For science writer and scholar Jon Turney, the book failed in its attempt at explanation because many questions went unanswered, but its strength was that it gave readers an impressionistic account of the workings of contemporary cosmology.[39] A publisher who turned down the book said: "My mistake was thinking that the reader would lose the arguments two-thirds of the way through and that this mattered."[40] The reason it did not matter, for novelist Gilbert Adair, was that Hawking's physical presence drove the book's success. If written by another physicist, he said, *A Brief History of Time* would have been an undoubted commercial failure.[41]

Other scholarly critics argued that Hawking wrote the book for a less obvious motive. He portrayed himself as a heavyweight in the history of science. He presented himself as a successor to Galileo, Newton, and Einstein. No clear reason existed for including the mini-biographies of these iconic scientists at the end of the book, wrote one scholar, except for the implication "that Hawking himself should be the next name on this list."[42] On this theme, historian of science Patricia Fara recounted a 1987 encounter with Hawking at Grantham, the English village where Newton went to school. At a conference to celebrate the great thinker's *Principia*, Hawking insisted he be photographed beneath the supposed descendant of the original, mythical apple tree that inspired Newton. Fara called it the "a bizarre re-enactment of a mythical event."[43]

THE PRIVATE LIFE OF A PUBLIC SCIENTIST

A Brief History of Time built a Hawking industry. Publishers raced to commission similar titles, believing that if one popular science book on an esoteric topic can become a best seller, then others can replicate its success, a trend described as the Hawking effect or the Hawking phenomenon.[44] Writers have produced two biographies. Kitty Ferguson focused on Hawking's scientific career and ideas.[45] John Gribbin and Michael White told his life story in terms of his rising fame, with chapter titles in the 2003 edition of the book such as "The Breakthrough Years," "The Foothills of Fame," and "Science Celebrity."[46]

Reviewing the biographies, critics examined the effect Hawking's newfound fame had on his private life. Hawking separated from his wife, Jane—and novelist Anthony Burgess argued that fame corroded the marriage. The money earned from *A Brief History of Time*, he wrote, "seems to have bought him the opportunity for a kind of emotional turbulence. He has left his wife, Jane, the true heroine of his story, sustained in her care of him and their children by a profound belief in God."[47] Scientist Bernard Carr, one of Hawking's former research students, chided White and Gribbin for near-sensationalism in their discussion of Hawking's marriage breakup. "The only people qualified to comment on this are the Hawkings themselves," wrote Carr, "and, since they have not done so, all that remains is gossip and speculation."[48]

Personal reflections on Hawking became part of other writers' books. Dennis Overbye, who profiled Hawking for *Omni*, made him a protagonist in his *Lonely Hearts of the Cosmos*, an account of modern cosmology and cosmologists. Overbye outlined Hawking's ideas, but also included his personal reflections about a physicist he found to be elusive yet eerily familiar. "Although I knew hardly anything about Hawking," recalled Overbye about his first seeing him in the flesh in 1976, "he struck me instantly as a charismatic figure. On some level that I didn't understand, I felt that I had always known about him."[49] Nervous before an interview with Hawking, he wrote: "I felt I was about to go on stage. I realized that I couldn't remember what Hawking looked like. I could imagine a slim figure in a wheelchair, but he had no face, only darkness."[50]

A disembodied Hawking was featured in the 1992 film version of *A Brief History of Time*. Directed by acclaimed documentarian Errol Morris, the film wove explanations of Hawking's physics with interviews with his mother and colleagues about Hawking. The film portrayed Hawking as connected fundamentally with the physics he studied. The opening section featured pictures of Hawking as a baby with his voice-over asking, "How real is time? Will it ever come to an end? Where does the difference between the past and the future come from?" His mother, Isobel, said, "Stephen always had a strong sense of wonder and I can see that the stars would draw him—and further than the stars."[51] The film's final image saw Hawking in his wheelchair floating into the starry cosmos. In production, Morris filmed Hawking before a blue screen so his image could be subsequently projected against any background. Morris said: "I can place Stephen Hawking where he belongs—in a mental landscape rather than a real one."[52] Hawking felt Morris had filmed him "like a sofa."[53]

In an article that publicized the film, *Vanity Fair* delved into Hawking's complicated private life. Journalist Arthur Lubow described how Hawking left his wife and moved into a nearby apartment in Cambridge with one of his nurses, Elaine Mason. *Vanity Fair* detailed the clash of personalities between Jane and Elaine. "Research students in Stephen's department thought Jane patronized him. She was constantly warning him of things he shouldn't do, treating him like an invalid. In contrast was Elaine Mason, the effervescent nurse who had immediately won Hawking's affection by refusing to treat him like an invalid." But the marriage breakdown affected the film, as producers were unable to secure interviews with Jane or their children. As the magazine put it, his public profile was now affected by "intrusions of messy personal affairs."

The piece uncovered Hawking's promotional apparatus. "The myth-making media machine has manufactured a Hawking icon of crippled genius that can be placed alongside the ear-bandaged van Gogh or the stone-deaf Beethoven," wrote Lubow. "Hawking is the model of pure mind unleashed from body." But he noted that journalists had a difficult task to get past this partly manufactured image. "Any journalist who has received a celebrity's rote response to an often asked question will flinch at this cybernetic nightmare," wrote Lubow. "Hawking literally serves up pre-cooked answers by accessing them from his computer memory." The film, the magazine concluded, would "pump another shot of adrenaline into Hawking's public reputation." Bantam also released at the time *Stephen Hawking's A Brief History of Time: A Reader's Companion. Vanity Fair* called it "a book of the movie of the book."[54]

Hawking indulged in the revelation of private details for commercial gain and public interest. His essay collection *Black Holes and Baby Universes*, first published in 1993, mixed pieces on theoretical physics with articles that revealed his earliest memory ("standing in the nursery of Byron House in Highgate and crying my head off"), a Christmas present of an American train set from his father ("I can still remember my excitement as I opened the box"), and his ALS diagnostic test ("They took a muscle sample from my arm, stuck electrodes in me, injected some radio-opaque fluid into my spine").[55] According to the promotional blurb, in the essays Hawking "is revealed variously as the scientist, the man, the concerned world citizen." Hawking the man sold well, spending three weeks—with number 14 as its highest placement—on the *New York Times* best seller list.[56]

Hawking became a brand, capable of selling not only his own books but also other cultural and commercial products. In 1993, he appeared in a ninety-second commercial for British Telecom,[57] had his voice sampled for Pink Floyd's song "Keep Talking," and starred in an episode of *Star Trek: The Next Generation*. Hawking appeared in a holographic poker game where his opponents were Einstein and Newton, a segment he would later replay at public relations events, including a visit to the White House. But in 1993 he denied he would write a sequel to *A Brief History of Time*. "What would I call it?" he asked. "*A Longer History of Time? Beyond the End of Time? Son of Time?*"[58] By the time of the quote in *Black Holes and Baby Universes*, Bantam had published *Stephen Hawking's A Brief History of Time: A Reader's Companion*. In 2005, it published *A Briefer History of Time*.

In 1995, Hawking married Elaine Mason. Newspapers worldwide reported the event. For the first time, volumes of media coverage focused exclusively on his private life. But the more analytical coverage used the event as an occasion to debunk the Hawking myth. The *Toronto Star* said Hawking's divorce and second marriage revealed "the emotional turmoil boiling beneath those wasted limbs and wearily lolling head. If nothing else, we see that the media image of Hawking as a cosmos-roving intellect freed of earthly desires and needs is misplaced."[59]

But it was not journalists who revealed the most about Hawking's private life. It was his former wife. For years publishers chased Jane Hawking to write a book. Her memoir, *Music to Move the Stars: A Life with Stephen*, first published in 1999, contained in its 612 pages intimate details of their courtship, marriage, and domestic and sexual life. "He required my help with the minutiae of every personal need, dressing and bathing, as well as with larger movements," she wrote. "He had to be lifted bodily in and out of the wheelchair, the car, the bath, the bed." The wider world never saw her former husband's daily struggles. "Outsiders could have no concept, though, of how painfully emaciated his body had become, like the corpses at Belsen which I once saw in a television documentary," she wrote. "Nor did outsiders generally witness those horrendous choking fits which would come on at supper-time and last well into the night, when I would cradle him in my arms like a frightened child till the wheezings subsided and his breathing slipped into the easy rhythm of sleep."[60]

One reviewer said the memoir was a typical kiss-and-tell book, written by a woman who married the famous and "glamorous" Hawking.[61] Jane Hawking said she had a different motivation. "Because Stephen is so famous, my life has become public property too," she told a reporter. "If I didn't tell my story, someone would come along in the future and invent it."[62] Notable is her belief that fame corroded their marriage. She told one reporter: "Fame and fortune muddied the waters and really took him way out of the orbit of our family."[63] For another reporter, family seemed inconsequential in the Hawking home "turned into a circus by his fame . . . nutty phone calls, the people arriving unannounced to lay on healing hands—and always the film crews."[64]

THE MEDIA UNMAKING OF STEPHEN HAWKING

By the mid-1990s, Hawking's media profile began to have a peculiar effect on his scientific reputation. A physicist argued that the creation of new scientific knowledge was being harmed by Hawking's cultural status. Unlike Newton or Einstein, argued Jeremy Dunning-Davies, Hawking's theories have not been verified. "However, criticism of these theories is, on occasion, restricted—even stifled," he wrote. "I know of colleagues who have had other papers rejected simply because the end result disagrees with Hawking."[65] Because Hawking's popular status is so high, "papers which challenge Hawking in purely scientific grounds are not successful because his reputation has in some sense gone beyond the purely scientific."[66]

At the end of the decade, journalists and his peers punctured the popular image of Hawking as one of history's great scientists. *Physics World*, the membership magazine for the Institute of Physics, polled in 1999 a sample of the world's leading physicists to name the five figures who made the most important contributions to the field. Only one of the 130 respondents put Hawking anywhere on their top-five list.[67] Physicists also disagreed with his prediction that there was a fifty-fifty chance a grand unified theory would be found before 2020. For example, Princeton University's Phil Anderson testified: "The question is an insult to me and to all those who call themselves theoretical physicists. A unified theory is unlikely to tell us much at all, though it may simplify a few questions about cosmology."[68]

The *Spectator* evaluated Hawking's scientific career. Robert Matthews, a trained physicist and science journalist, examined what he saw as the three significant contributions Hawking had, by then, made to physics. For Matthews, Hawking's postdoctoral work on singularities was "an exercise in academic loose-end tying: important, but hardly epochal." Hawking radiation is accepted as true, but there is no evidence to prove the theory. And evidence that the universe will expand forever refuted the idea that space and time will collapse in a big crunch. But these achievements were exaggerated as journalists constructed Hawking as a world-beating scientist, a process that revealed "the foibles of the media when dealing with things it can't quite get its head around." He added: "For unlike his fellow scientists—or, come to that, the man himself—the media continues to cling to the tabloid notion of Hawking as a genius trapped in a useless body."[69] On the same theme, British astrophysicist Peter Coles in *Hawking and the Mind of God* put Hawking's science in perspective. Hawking was a significant theoretician of his time, he wrote, but to compare him to Einstein and Newton was "absurd."[70]

Reporters noted how Hawking's name, at times, became little more than a marketing device. Reviewing the six-part PBS exploration of the cosmos—*Stephen Hawking's Universe*—one reviewer said the physicist "tops and tails each episode and chimes in with the occasional comment, but he is mainly used here for name value."[71] The *New York Times* praised the show's faithfulness to its material, but noted that the companion book to the series was not written by Hawking. The scientist, wrote the paper's reviewer, was described as "the series host, a word that on television often means a personality who lends star power to the creations of others."[72] In another way, influential science journalist John Horgan characterized Hawking's science *itself* as promotional—for astrophysics. In *The End of Science* he said Hawking's cosmology was not science in the strictest sense because its theories were not testable or resolvable. "Its primary function," wrote Horgan, "is to keep us awestruck before the mystery of the cosmos."[73]

Commentators also scrutinized the Hawking industry after the publication of *The Universe in a Nutshell.* Jon Turney in a review concluded that because physics had not advanced significantly since 1988, there was little need for the new book—"apart from Bantam's urge to keep the franchise going."[74] John Gribbin, a biographer of Hawking, raised serious questions about the degree of authorship Hawking exerted over the book. "Whereas previously the party line was that A Brief History was all his own work (which anyone who has seen the early draft knows cannot be true), this time Hawking thanks Ann Harris and Kitty Ferguson, 'who edited the manuscript.'"

"I don't know the work of Ann Harris," continued Gribbin, "but some of the less technical parts of the present book certainly read like the work of Kitty Ferguson." Unless Hawking has started to believe his own hype, the reviewer argued, he should have told his publishers to be more modest in the book's blurb. "'Great' is an adjective that should be used sparingly, and when used in science reserved for the likes of Albert Einstein and Richard Feynman." He added: "Someone at Bantam should have the guts to tell Hawking that his jokes aren't funny."[75]

Other physicists' grumblings spilled into newspaper reports. "Coffee-time talks in physics departments often come up with the same topic: it's very difficult to get anybody to say anything critical of him," said Peter Coles, professor of astronomy at the University of Nottingham, in a piece in the *Independent.* "But to have somebody like that in an establishment that runs on peer review isn't healthy. The trouble is, people fear that they will be thought of as jealous."[76] A cosmologist anonymously told the *Independent*: "To criticize Hawking is a bit like criticizing Princess Diana—you just don't do it in public."[77] Neil deGrasse Tyson recollected in his autobiography that late-night talk at physics conferences inevitably slid into a debate about whether Hawking's scientific work warranted his popular status.[78]

One physicist who criticized Hawking in public was Professor Peter Higgs, who gave his name to the Higgs boson, the so-called God particle. He argued that Hawking's fame gave enhanced scientific merit to his theories. "It is very difficult to engage him in discussion," the *Times* reported Higgs as saying, "and so he had got away with pronouncements that other people would not. His celebrity status gives him instant credibility."[79]

When Hawking published *On the Shoulders of Giants* (2002), an edited collection of foundational scientific texts from Copernicus, Galileo, Kepler, Newton, and Einstein, his historical expertise was ridiculed. In a *Nature* review, Owen Gingerich, a professor of astronomy and the history of science, said Hawking's historical writing was filled with errors. Copernicus was definitely not, as Hawking asserted, a priest. "Should I be embarrassed for Stephen Hawking because an enterprising publisher has inveigled him into putting his name to a collection of superseded texts?" asked Gingerich. "Or should I be outraged that an eminent scientist, but one with no track record in the history of science, has the arrogance to endorse historical introductions for five classics of science?"[80]

Popular media pumped up Hawking's image as an iconic scientist. Now popular media punctured what many considered an inflated scientific status.

MANY HAWKINGS COLLIDE

By the early 2000s, several narratives about Hawking circulated through popular culture. He was the disembodied genius, the embodied father and husband, the scientific great, and the overhyped star. These narratives collided in 2004.

News organizations worldwide reported claims that, since 2000, Hawking had been hospitalized with unexplained injuries, including a broken wrist, cut lip, and gashes to his face. Police questioned him and his wife Elaine over allegations that he was abused. The Melbourne-based *Sunday Age* wrote in a headline: "Just What Is Going On in the Hawking Home?"[81] Journalists reported that, in the wake of the allegations, his daughter, Lucy Hawking, sought treatment for depression and alcoholism.[82] Police stopped their inquiry after they found no evidence to substantiate the claims; Hawking rejected publicly claims he had been abused. His son Tim was reported as saying: "I'm very concerned. He denies it every time I speak to him about it and I would hope that he would respect me enough to tell me the truth."[83]

The couple clumsily presented a united front for the media. "Stephen Hawking's second wife Elaine staged an astonishing Valentine's Day stunt in response to those who accuse her of cruelty towards the scientist," reported the *Daily Mail*. "She tied a red, heart-shaped balloon emblazoned with the words 'I love you' to his motorised wheelchair then took him to lunch at a Thai restaurant near their home in the centre of Cambridge. After the meal, with the balloon still flapping in the wind, they joined other couples strolling along the banks of the Cam."[84]

In the same year his public and private worlds collided in the television drama *Hawking*. Produced by the BBC, the show focused on two crucial years of his postgraduate study when he was first diagnosed with ALS, met Jane Wilde, and worked on singularities for his doctoral thesis. Made with his cooperation, Hawking complained that the first draft of the script did not focus enough on scientific issues and was what he called a "soap opera."[85]

In 2004 his science was again opened to public scrutiny. He asked to speak at a major scientific conference, promising that he had solved a four-decade-old problem in theoretical physics—the black hole information paradox. His 1970s work showed heat leaked out of a black hole, leading to its explosive end, a demise that would also destroy all information about the cosmic matter, such as stars, that the black hole had swallowed. But scientists knew as an iron-cast physical law that information could never be lost. So where did the information go? Organizers did not know Hawking's solution, because, unlike other speakers, he did not submit a paper or summary beforehand. "To be quite honest," physicist Curt Cutler told *New Scientist*, "I went on Hawking's reputation."[86] Flashbulbs flickered as Hawking approached the podium to admit before more than six hundred of his peers—and the world's media—that he was wrong to say a black hole destroyed information it swallowed. But he also suggested that the lost information traveled through the black hole to a parallel universe. Yet many physicists were not convinced. They reserved judgment until Hawking laid out his idea in a paper with detailed mathematics.[87]

Hawking's preparation of this mathematical proof became the storyline of a 2005 BBC documentary, *The Hawking Paradox*. Its director, William Hicklin, wanted to tell a complex tale of Hawking's attempts to solve a perplexing problem first posed in the 1970s. But he believed a story about a technical piece of high-end physics would be too complicated for a mass audience. Hicklin, instead, told a more dramatic and understandable story of Hawking's quest to restore his status among his peers, many of whom believed he had done little significant work since his pioneering 1970s research. "I suppose the idea of the film," Hicklin said, "was that at the end of his life he was trying to regain his reputation within the physics community."[88] The director said Hawking promised the proof would be published before production ended.

Hawking worked with the producers to shape his on-screen image. He answered questions himself, indicated his preferred camera angles, and allowed himself to be filmed in a white studio where his image could be superimposed on different backgrounds. *The Hawking Paradox* repeated standard ways of constructing Hawking, discussing how he works on problems in his mind inside his disabled body, his discovery of Hawking radiation, his unique way of doing physics. For science studies scholar Hélène Mialet, Hawking's input helped the film standardize his legend as a disembodied genius physicist.[89] Hawking did not finish his paper on the information paradox by the time shooting finished. It was published in October 2005. Physicists agreed Hawking had not solved the information paradox.[90]

Physicists further criticized publicly Hawking's research and reputation. When he received the Royal Society's Copley Medal for science communication, an eminent scientist said his promotion of grandiose scientific theories presented a misleading view of day-to-day scientific work. Hawking, the scientist anonymously told the *Sunday Times*, was "a great embarrassment to us."[91] After Hawking publicly stated that it would be more exciting if researchers *failed* to find the God particle—the elusive particle important for understanding the subatomic world—Peter Higgs criticized his colleague's research. Hawking's work on particle physics and gravity, he said, was "not good enough."[92]

Even though different facets of Hawking's life were laid out, journalists found it difficult to write about Hawking because his essential nature remained elusive. For one journalist, the difficulty lay "in the stuff that you cannot understand, that you cannot . . . REACH. Mystery swirls about him like mist over a bottomless quarry."[93]

GODMONGERING AS MARKETING

Hawking continued to explain to general audiences the up-to-the-minute ideas in physics. Coauthored with Caltech physicist and writer Leonard Mlodinow,

The Grand Design (2010) argued that M-theory, a grand mathematical framework that unified various theories of the basic structure of the universe, was now the best candidate for a theory of everything. The book described the multiverse, perhaps the most important concept in contemporary physics, the speculative and unproven idea that a number of universes are constantly created, each with its own physical laws. Other physicists popularized the idea, but as *Nature* said in its review: "when Hawking speaks, people listen."[94]

Before explaining the multiverse, Hawking and Mlodinow wrote off philosophy and religion. Philosophy was dead because the field was disconnected from physics. God was no longer needed to explain the beginning of the universe,[95] a claim that inevitably led to blanket media coverage: "Universe not created by God, says Hawking," wrote the *Guardian*.[96] "Has Hawking Seen Off God?" asked the *Telegraph*.[97] The announcement was greeted, wrote science writer Philip Ball, as though it were "the final judgement of science on the Biblical creation: Hawking Has Spoken."[98]

Experts—including Hawking's peers and colleagues—excoriated the book. Physicist Roger Penrose said the book's ideas were so removed from testable verification that they were "hardly science."[99] Astronomer Martin Rees said Hawking has read so little theology that readers should not attach any weight to his views.[100] A philosopher of physics said that after the authors claimed philosophy was dead they went on to "unwittingly develop a theory familiar to philosophers since the 1980s, namely 'perspectivalism,'" a theory asserting that "science offers many incomplete windows onto a common reality, one no more 'true' than another."[101] Graham Farmelo, a biographer of Paul Dirac, said the book felt like a padded-out magazine article. "It gives me no pleasure at all," he wrote, "to say that I doubt whether *The Grand Design* would have been published if Hawking's name were not on the cover."[102]

The Grand Design's superficial religious arguments, critics argued, were made to attract publicity. The invocation of God was an example, for the *New York Times*, of Godmongering: the use of religious issues as attention-grabbing headlines. The paper said Hawking was "a formidable mathematician and a formidable salesman."[103] Indeed Hawking Godmongered for publicity, telling ABC News, for example, "The laws of physics can explain the universe without the need for a creator."[104] *Nature* began its (favorable) review: "Despite publicity to the contrary, *The Grand Design* does not disprove the existence of God."[105] The *Guardian* lampooned the Godmongering strategy. "[T]hanks, Stephen, that's lovely," concluded a satirical article on the book. "If you could just end with something you haven't written before to create a few headlines, then we're done. How about God doesn't exist? Lovely job. Let's do it all again in a couple of years."[106] Hawking actually did so a year later, telling the *Guardian* in an exclusive interview to promote a lecture: "There is no heaven or afterlife . . . that is a fairy story for people afraid of the dark."[107] The Godmongering apparently worked: the book debuted at number 1 on the *New York Times* nonfiction best seller list, and stayed on the list for several weeks.

BUILDING THE HAWKING LEGACY

The Hawking industry intensified in 2012 to mark his seventieth birthday. Kitty Ferguson, who wrote an early biography of Hawking, produced *Stephen Hawking: An Unfettered Mind*, written with his cooperation. She defended Hawking against his detractors. On the questions about his level of input into *The Universe in a Nutshell*, she told how Ann Harris—Hawking's editor at Bantam—first sent her a disorganized collection of Hawking's public and scientific writings, and asked if Ferguson could assemble the material into a book. "Stephen Hawking was one of the jewels in her and Bantam's crown," wrote Ferguson. "It was unthinkable to send this back to him and say it couldn't be published."[108]

On the scathing criticism of *The Grand Design*, Ferguson claimed several commentators seemed not to have read the book.[109] On the marriage to Elaine and the abuse allegations, she called the period a "difficult-to-understand chapter in [the Hawkings'] lives."[110] On criticisms about his theological claims, she said his quotations can be used to support those who believe in God and those who do not—he's "been the hero and villain of both camps."[111] On Hawking's self-promotion, she noted that by 1995, "Hawking had become a master manipulator of his public . . . as one of his personal assistants once commented to me, 'He isn't stupid, you know.'"[112] Reviews were respectful, but at least one critic was tired of reading the bland Official Version of Hawking. Ed Lake in the *Daily Telegraph* said Ferguson's "starry-eyed" treatment had "so little that's dark or sad" about Hawking that "the effect is almost sinister," as if something was being left out. "One looks forward to the kind of muckraking biography his celebrity cries out for."[113]

Hawking was celebrated in other ways. As the chart indicates, in 2012 media coverage of Hawking peaked.[114]

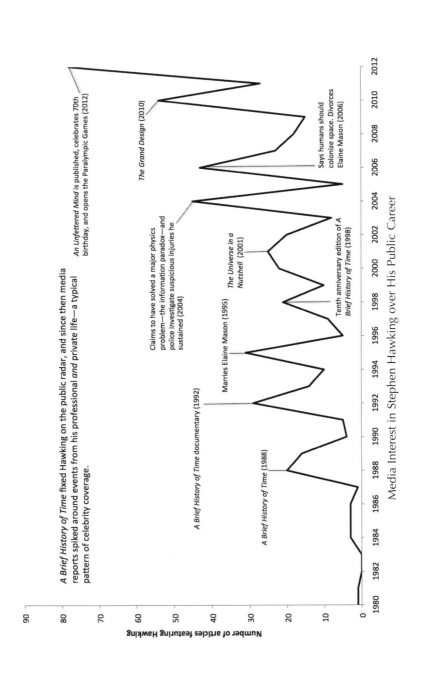

A Brief History of Time fixed Hawking on the public radar, and since then media reports spiked around events from his professional *and* private life—a typical pattern of celebrity coverage.

An Unfettered Mind is published, celebrates 70th birthday, and opens the Paralympic Games (2012)

The Grand Design (2010)

Claims to have solved a major physics problem—the information paradox—and police investigate suspicious injuries he sustained (2004)

The Universe in a Nutshell (2001)

Marries Elaine Mason (1995)

Says humans should colonize space. Divorces Elaine Mason (2006)

Tenth anniversary edition of A Brief History of Time (1998)

A Brief History of Time documentary (1992)

A Brief History of Time (1988)

Number of articles featuring Hawking

90 80 70 60 50 40 30 20 10

1980 1982 1984 1986 1988 1990 1992 1994 1996 1998 2000 2002 2004 2006 2008 2010 2012

Media Interest in Stephen Hawking over His Public Career

London's Science Museum opened a temporary exhibition of his life and work, featuring new photographs of him in his office, a little-seen portrait by artist David Hockney, the annotated script of his appearance on *The Simpsons*, and the blue flight suit he wore for a 2007 zero-gravity space flight. He was the star turn at the cosmological-themed opening of the Paralympics, as he was symbolic of the triumph of the human spirit, described by the organizers as "the most famous disabled person in the world."[115] At Cambridge, a carefully forged statue of Hawking will stand in the gardens of the Centre for Mathematical Science. A Permanent Hawking Archive at the university will house his scientific papers, his books, his press coverage, and a typescript of *A Brief History of Time*. The Hawking legacy is built to endure.

Hawking contributed to his end-of-career legacy building with his 127-page autobiography *My Brief History*. As well as including new personal pictures, he portrayed his second wife in a good light. "My marriage to Elaine was passionate and tempestuous. We had our ups and downs, but Elaine's being a nurse saved my life on many occasions." One such incident occurred after Cambridge recruited him to fundraise for its 800th anniversary. "I was sent to San Francisco, where I gave five lectures in six days and got very tired," he wrote. "One morning I passed out when I was taken off the ventilator," he recalled. "I would have died had not another caregiver summoned Elaine, who resuscitated me. All these crises took their emotional toll on Elaine."[116]

In the book, Hawking emphasized his scientific accomplishments. He wrote that most physicists would agree he was right about Hawking radiation, "though it has not so far earned me a Nobel Prize,"[117] because its theoretical concepts are very difficult to prove with experimental evidence. "On the other hand," he noted in the next sentence, "I won the even more valuable Fundamental Physics Prize."

Despite the revelations about his marriages, the criticism of his career, his comments on his relationship, his defensiveness of his lack of a Nobel Prize—all elements that show him to be a real human—one researcher argued that the public image that dominated was not that of Hawking the man. It was instead a construct created and sustained by a massive support network and a largely uncritical media—a construct called HAWKING.[118]

A photograph taken in 2013 visualizes this idea of a real and a represented Hawking. During a talk by the physicist at London's science museum, he projected behind him a giant image of the cover of the first U.S. edition of *A Brief History of Time*. The cover, which Hawking said he had no control over, has come to be his signature public image—a mind floating in space.

Hawking speaking at London's science museum in 2013 before a picture of the first U.S. edition of the book that propelled him to global celebrity. Its cover illustration helped create his distinctive image as a disembodied mind floating in space to discover pure knowledge. Andrew Matthews/PA Wire URN:18185512 (Press Association via AP images)

THE MOST FAMOUS CELEBRITY SCIENTIST

Hawking's status as the most famous scientist of modern times is grounded in the quality of his research. In the 1960s and 1970s, Hawking worked at the forefront of cosmology as the field produced revolutionary new understandings of our universe. He made a lasting contribution to this new physics with his 1974 idea that information leaked out of black holes. The science came first.

Then magazine writers framed him as the personification of the era's strange new physics. They mixed his personal and private lives to present him as an idiosyncratic scientist whose mind was somehow set free from a trapped body to roam in a mysterious cosmic realm and discover deep truths of the universe. Hawking's symbolism was set, even though early coverage, paradoxically, showed that, far from roaming free in the universe, he relied on a network of Earth-bound colleagues, collaborators, and caregivers to help him work.

Nevertheless, the physicist's image and science were packaged, marketed, and sold in *A Brief History of Time*, a book that spread his ideas through society and staked out his lasting place in popular culture and, perhaps, scientific history. Yet marketing alone cannot account for the book's unprecedented success. Hawking and his account of modern physics connected in a deep way with end-of-the-century audiences hungry for answers to the origin and meaning of humankind's existence.

Hawking's post–*A Brief History of Time* public career illustrates the less welcome but inevitable consequences of fame. Hawking's private life was quickly opened to public scrutiny, often for prurience and profit. Critics eviscerated his scientific reputation, especially since all agreed none of his subsequent research matched the impact of his early black hole research. This led to a peculiar situation where he is celebrated in the public eye as a scientific great, but he is not rated by his peers as among the top physicists of all time. That is another paradox of his fame.

But overriding all these elements is the crucial feature of Hawking's stardom—his unique ability to *symbolize* an idea of science as the cerebral search for pure truth. He became a modern incarnation of a figure hardwired into the history of Western culture: the disembodied mind. Our celebrity culture saw in Hawking the idea made vivid and real.

THREE

Richard Dawkins's Image Problem

During a photo shoot for the *Observer* in 2013 to promote the first volume of his memoirs, Richard Dawkins asked to see the photographer's pictures. The photojournalist, Andy Hall, showed him the screen of his digital camera. The newspaper reported what happened next.

> Dawkins told Hall: "You've made me look too harsh."
> The photographer said he wanted to portray the scientist's gravitas.
> Dawkins replied: "I don't want f***ing gravitas . . . I want humanity."[1]

By printing the exchange, the newspaper showed readers the struggle that underlies the best celebrity interviews: the contest to control public image. The star—and by this time Dawkins was undoubtedly a star, the world-renowned atheist and author of a succession of global best sellers—wants to project a favorable image of him- or herself, usually to promote the latest film, book, or other venture. The reporter and photographer, however, try to penetrate this public relations veneer to reveal the unvarnished essence of the star's personality.

The paper interviewed Dawkins as he promoted *An Appetite for Wonder: The Making of a Scientist*, an intimate account of his childhood in colonial Africa and his early career as a biologist at Oxford. The book's promotional description printed on the inside of the cover argued that *The God Delusion*, his 2006 polemic against religion, crystallized Dawkins's public image into that of a brilliant but strident proponent of "ruthless skepticism." The memoir, in contrast, offered "a more personal view."[2] The *Observer* journalist wrote that Dawkins seemed "determined in both the memoir and our interview to present a calm, conciliatory side to his character that has not always been associated with his public image."

A main theme of the article was Dawkins's attempt to shift his public image. The reporter fairly portrayed Dawkins as a passionate advocate for secular humanism. But he also reported the scientist's contentious remarks, noting that after Dawkins described what he considered the weaknesses of Muslim science, his "controversy seismometer" went off the Richter scale. Yet the journalist concluded: "we part company on a benign note, without controversy or dispute, the great humanist having succeeded in showing his humanity."

This idea was reflected in the striking photographic portrait that illustrated the article. Hall, the photographer, framed Dawkins against a blue sky, a shot traditionally chosen to present the subject as a hero. Dawkins

Richard Dawkins, photographed by the *Observer* in 2013 as he promoted his autobiography in which he presented a more intimate view of himself to counter what one reporter called "the strident, abrasive caricature beloved of lazy journalists." Photo by Andy Hall/Contour by Getty Images

gazes down at the viewer with a tight-lipped smile, behind him the sun breaking through scattered clouds, a shard of reason's light in his feared world of irrational darkness.

The fight to define Dawkins's image has implications far beyond selling books. Dawkins's public career shows the power of celebrity to mold a talented researcher into perhaps the most influential scientific public intellectual of our time. While Stephen Hawking remained at the heart of elite physics while he wrote his popular books, Dawkins moved further and further away from academic science throughout his career, using his scientific celebrity to forge a new identity as the leader of a new social movement of atheists.

As a celebrity public intellectual, Dawkins came to stand for such vital scientific and cultural ideas as the gene's-eye view of evolution, the power of Darwinism to explain life, the central role of science and reason in personal and political affairs, and the social need for proud atheism. He gave these complex and contentious ideas a human face, making these abstractions visceral and real as he became "Professor Evolution" and "Professor Science"[3] and "Professor of Atheism."[4] How he is portrayed therefore has consequences for how these ideas are viewed and understood by citizens. Dawkins's public image matters for the public image of science.

THE GENE MACHINE

In the 1960s and 1970s, Dawkins was a rising star in biology. He studied zoology at Oxford, a powerhouse of evolutionary thought, a center of the adaptationist school that stressed how natural selection modified living things so that they would be better suited to survive and reproduce. Dawkins specialized in ethology, the biological study of how animals behave, usually in their natural environment, but also in a laboratory. Dawkins worked as a research assistant to Niko Tinbergen, a Nobel Prize–winning founder of the field. For his doctoral thesis, Dawkins studied the pecking behaviors of baby chicks to see, among other things, if this habit was ingrained from their birth.

A gifted researcher and communicator, his first conference presentation on his work was so impressive that he was recruited for a job at the University of California, Berkeley. As well as teaching zoology there, he participated in anti-Vietnam protests and pro-Democratic rallies before he returned to Oxford. He had published in scientific journals since 1968 and for periods in the 1970s served as European editor of the journals *Animal Behaviour*

Monographs and *Animal Behaviour*. In 1970 he was appointed as university lecturer in zoology and was well on the path to a scientific research career.

Ethology interested the public. Zoologists such as Konrad Lorenz and Desmond Morris wrote best sellers that linked studies of animal behavior to evolution, telling readers how similar people were to beasts in their sexual and professional lives, a narrative lapped up by general readers (not least because it provided a biological excuse for uncivilized behaviors).[5] But several of these popularizations frustrated Dawkins, as he believed they spread the mistaken view that natural selection worked to ensure the survival of the species.

Dawkins disagreed. He sided with evolutionary biologists at Oxford, such as William Hamilton and John Maynard Smith, who were part of a revolution in biology that tied the new findings of genetics to the older ideas of Charles Darwin to show *how* natural selection worked. These neo-Darwinists argued that natural selection did not work to ensure the survival of the species or individual. It worked to ensure the survival of the gene. This was the gene's-eye view of evolution.

The idea electrified Dawkins. He had lectured on the gene's-eye view of evolution and he longed to spread these ideas wider. At the time, he was immersed in a research project on the behavior of crickets. But in 1973 his laboratory work was interrupted after a strike by coal miners led to electricity rationing to save fuel. With only enough power for a three-day working week, Dawkins postponed his research, sat down with a portable typewriter, and produced the first chapter of a book that would explain the gene's-eye view of evolution to the public.

The strike ended and the power returned. Dawkins consigned the manuscript to his desk drawer where it remained for two years. As more papers piled up on the gene's-eye view, he returned to the project with renewed vigor, later recalling that he wrote it "in a frenzy of creative energy." When Michael Rodgers of Oxford University Press read his work, he called Dawkins and said in his loud voice: "I MUST HAVE THAT BOOK!" The press committed to publish it in a record six months. He and the publishers considered calling it *The Immortal Gene* or *The Cooperative Gene* or *The Altruistic Vehicle* or *The Gene Machine*.[6] Dawkins called it *The Selfish Gene*.

Oxford University Press predicted the book would have a public impact—because it would contribute to a scientific controversy that had just flared up. In 1975, the Harvard biologist E. O. Wilson generated fierce controversy with his book *Sociobiology*. He examined the biological basis of animals' social behavior and concluded that aggressiveness, morality, religious beliefs, and sex roles were connected to our evolutionary past. After World War II, the idea that biology explained behavior was taboo. The Holocaust was built on the abhorrent view that some races were innately inferior. Decades of civil rights action in the United States had worked to overturn racist claims that blacks were naturally subordinate.

Wilson's book was so incendiary that a collective of left-wing and Marxist-influenced scientists, the Sociobiology Study Group (SSG), formed to challenge its claims. With evolutionary biologists Stephen Jay Gould and Richard Lewontin as its most visible members, the group argued that *Sociobiology* had a political message: they interpreted it to mean that because some people are destined to be second-class because of their biology, why bother to engineer society to make people equal? Sociobiology, in their view, justified unequal distributions of power in society, provided scientific legitimacy to the status quo, and discouraged social reform.

In this climate, the publisher Michael Rodgers at Oxford told his colleagues that sociobiology was "red hot"—and this could only help sell *The Selfish Gene*.[7] The book was first published in 1977. The month the book was published, the BBC's flagship science show, *Horizon*, broadcast an episode titled "The Selfish Gene." Dawkins was interviewed, but the host was his mentor and doyen of evolutionary biology, John Maynard Smith.

Dawkins stayed away from political or social commentary. He focused on the difficult task of explaining the gene's-eye view for general readers—without mathematics. He instead used brilliant metaphors. Genes were selfish in that they tended to increase their chances of reproduction and survival compared to other genes. Humans were survival machines for their genes in that humans died, but their genes lived on. He concluded the book with the optimistic humanistic sentiment that, unlike other animals, humans were not slaves to their selfish genes. They were driven also by morality and ethics. (The *New York Times* called it an "almost Hollywood-style ending."[8])

The book received generally favorable reviews. The *New York Times* said the book was the first popular treatment of neo-Darwinism and that it set a high standard for future accounts. The *Washington Post* reviewer said the book went beyond science to offer a bleak but compelling view of the human condition. It was "one of the coldest, most inhuman and disorienting views of human beings I have ever heard," he wrote, "and yet I love it! It is so deep an insight, to bridge the gap between the lifeless and the living, the chemical and biological, the random and the teleological, the physical and the spiritual." *The Selfish Gene* provided for the reviewer a "lucid and eerie view of who we are."[9]

Dawkins became swept up in the sociobiology controversy.[10] A 3,000-word *Newsweek* feature in 1978, for example, discussed Dawkins's ideas in a piece on sociobiology headlined "Our Selfish Genes."[11] The book also had a particular political resonance in Britain. It was interpreted by left-wing critics as providing a biological justification for the conservative policies of Prime Minister Margaret Thatcher. Her politics of individualism on a social level, they argued, mirrored selfishness on a genetic level.[12] Psychologist Oliver James said the book's idea captured the *Zeitgeist* of the greed-is-good individualism of the 1980s.[13] Philosopher Mary Midgely linked Dawkins to these cultural currents in a review, calling him an "uncritical philosophical egoist" whose typical readers were "people with vaguely egoist leanings about individual human psychology." His "crude, cheap, blurred genetics," she wrote in the journal *Philosophy*, "is the kingpin of his crude, cheap, blurred psychology."[14]

Dawkins presented the book as original scientific knowledge. "Rather than propose a new theory or unearth a new fact," he wrote, "often the most important contribution a scientist can make is to discover a new way of seeing old theories or facts."[15] The book influenced professional biologists, as it provided a clear modern Darwinian framework for their research. "I am convinced," wrote evolutionary biologist Alan Grafen, "that *The Selfish Gene* brought about a silent and almost immediate revolution in biology." The influence was silent, he argued, as it was not respectable for scientists to cite the book, because it did not present its ideas in the mathematical form expected by population geneticists, who had authority within evolution.[16]

Dawkins wrote his next book, *The Extended Phenotype* (1982), for expert evolutionists. He extended the selfish gene into a more speculative concept: that genes can increase their survival chances by reaching out of their organisms' bodies to have impacts—phenotypic effects—on the broader world. For example, a beaver's dam and its lake are as much the product of genes as the beaver itself: the lake protects against predators and provides access to food. The genes associated with well-built dams therefore survive at cost of rival genes associated with poor dams.

Dawkins viewed the book as his capital piece of original research: "The last four chapters," he wrote, "constitute the best candidate for the title 'innovative' that I have to offer."[17] The *Quarterly Review of Biology*—in a review headlined "Son of Selfish Gene"—concluded that there is "little new in it for population geneticists."[18] *American Scientist* gave it cautious praise, recommending it for readers as a "readable but very personal account of the evolutionary process."[19] *Science* magazine was not entirely convinced by his thesis, but said the book showed why evolution was so exciting in the early 1980s.[20]

THE ELOQUENT CHALLENGER
TO CREATIONISM

Creationism chiseled at this excitement. The movement rose as a cultural force in the 1970s and 1980s after a remodeled Republican Party under Ronald Reagan merged business and religious conservatives under one political banner. Creationists became self-styled creation scientists, arguing that the fossil record was formed by Noah's flood and the world was forged between 6,000 and 10,000 years ago. (By the late 1990s, creationism morphed into intelligent design. Its advocates argued life forms were so complex they must have been designed by a supernatural creator. They claimed unresolved issues about the mechanics of evolution meant the theory had major weaknesses and lobbied for public schools to teach this supposed controversy over evolution.)

Creationists made their case in public debate. The Institute for Creation Research (ICR), the movement's flagship organization, made university debates a cornerstone of its promotional activity. In the 1970s and 1980s, it conducted more than 300 public discussions across America.[21] Their views had significant public support. In 1982, a Gallup opinion poll found 44 percent of Americans surveyed believed God created humans in their present form, 47 percent believed in the theory of evolution—yet 38 out of that 47 percent believed God directed evolution, and 9 percent believed evolution occurred without God.[22]

Against this backdrop, in 1986 Dawkins published *The Blind Watchmaker*. "Darwinism seems more in need of advocacy than similarly established truths in other branches of science," wrote Dawkins. And he took on that role as public advocate, framing the book as a direct challenge to creationism, giving it the subtitle: *Why the Evidence of Evolution Reveals a Universe without Design*. He demonstrated how gradual evolution shaped all life on Earth, using the sonar of bats and the evolution of the human eye to illustrate that cumulative natural selection, over time, was nonrandom and had no long-term aim, no final destination, no goal except short-term survival or reproductive success. He wanted the book to "persuade the reader, not just that the Darwinian world-view *happens* to be true, but that it is the only known theory that *could*, in principle, solve the mystery of our existence."[23]

The Blind Watchmaker spread these ideas through culture. Its paperback edition reached number 2 on the *Sunday Times* best seller list in 1988, and that year the *New York Times* listed it as a notable paperback. Dawkins presented a *Horizon* show on the book's central theme. The *Sydney Morning Herald* labeled the book the "seminal work on the contemporary state of the evolution-versus-creationism debate."[24] The *New York Times* said it "may provide assistance for those attempting to defend science from sectarian attacks."[25] The *Times* of London framed the book as "an extended polemic for Darwinism."[26]

Darwinism flourished in 1990s academic life and popular culture. It migrated from evolutionary biology to other fields, creating new ways of understanding older disciplines and in some cases forming new fields, such as evolutionary psychology, evolutionary economics, evolutionary medicine, and even evolutionary literary criticism. Darwinism was, in the memorable phrase of philosopher Daniel Dennett, a universal acid that corroded and burned through whatever it touched.

Darwinism captured the public imagination, as an explosion of popular titles explained natural selection but also promised readers insights into the origin and meaning of life.[27] "Richard Dawkins," wrote British cultural critic Melvyn Bragg in 1998, "has surfed with spectacular success on the triumphant tide of modern biology, which has proved Darwinism in all its oceanic reach to be the 'ism' of our time."[28] Dawkins expanded on this theme in two 1990s books, *River out of Eden* (1995) and *Climbing Mount Improbable* (1996), which both discussed the overarching theme of his public career: the almost limitless power of Darwinism to explain life.[29]

Dawkins and Gould were popular Darwinism's two most influential writers. Each personified contrasting schools of evolutionary thinking. Gould and Dawkins disagreed over the mechanics of evolution and the importance of natural selection. Dawkins argued that natural selection operated on the gene and that evolution occurred gradually through the complex adaptation of gene lineages over long time periods. He and his colleagues have been called gradualists or adaptationists. They believed that evolution could be examined through competition and adaptation between genes. Gould has looked at the large-scale history of the evolution of life on a paleontological timeline, believing natural selection also acted on the levels of organisms, groups, and species.

Around them formed a matrix of scientists and philosophers who wrote about Darwinism. Journalist Andrew Brown in *The Darwin Wars* (1995) divided the writers into either Gouldians or Dawkinsians. He said the clash between them sparked a "revolution in popular culture."[30] Their conflict spun off its own spin offs: *Dawkins vs. Gould: Survival of the Fittest* (2001) explained for popular audiences the dramatic clash of ideas. The *Observer* in 1998 described the conflict as "Survival of the Bitchiest." Both scientists, the paper said, were "suave, self-assured self-promoters." The academic trash-talk was sad, said the paper, because both had been instrumental in the public understanding of evolution. "On the other hand," the paper noted, "there is nothing like a nice petulant bout of insults to sell books."[31] As historian of biology Ullica Segerstråle wrote: "Controversy pays, and . . . it takes two to tango."[32]

Together Dawkins and Gould stood out as easily the most prominent enemies of creationism. As a result, Phillip Johnson, a Presbyterian lawyer who taught at the University of California at Berkeley and an influential figure in the Intelligent Design movement, once said he wanted to debate Dawkins or Gould as they were the significant public representatives of Darwinism. By the middle of the 1990s, the science writer John Gribbin called Dawkins "Darwin's new bulldog."[33]

THE HAWKISHLY HANDSOME
SPOKESMAN FOR SCIENCE

Dawkins was by the mid-1990s an established scientific star. The *Guardian*, for example, reported from a 1996 literary festival in Brighton, close to London, where a 6 PM lecture by Dawkins was sold out: "Richard Dawkins is nervous. This is very surprising," and noted that "he has a number of confidence-bolstering things going for him: a brain the size of a small planet, a command of language that would make your average novelist squirm with envy, a chair at Oxford, and the fierce, hawkish good looks of a forties film star."

The audience, the journalist noted, contained Dawkins's third wife, actress Lalla Ward, well known for her role in the sci-fi show *Doctor Who*. She now also illustrates his books and created the evolution-inspired designs for his ties. Dawkins captivated the crowd. "When he speaks, the whole theatre strains forward to listen. When he stops talking, people seem to relax a little, as if to help them begin digesting the platefuls of pertinence washed down by beakers of brilliance," wrote the *Guardian*. "When it is over, the applause explodes, and goes on until Dawkins leaves the stage."

The *Guardian* article also illustrated the popularity of science writing. Dawkins was the warm-up to the evening's literary event, in which novelists were to discuss spy thrillers. But that was canceled. Not one ticket was sold.[34]

This public prominence and cultural influence that Dawkins enjoyed was codified into an academic position in 1995. He resigned his prestigious Readership in Zoology to become the inaugural Professor of the Public Understanding of Science at Oxford. It was a position created for a scientist who could convey to the public the excitement of doing science. Dawkins became, in the words of philosopher Michael Ruse, an "anointed spokesman for science."[35]

For science journalist John Horgan, Dawkins was a perfectly calibrated public intellectual. "He is an icily handsome man, with predatory eyes, a knife-thin nose, and incongruously rosy cheeks. He wore what appeared to be an expensive, custom-made suit," Horgan wrote in *The End of Science* (1996). "When he held out his finely veined hands to make a point, they quivered slightly. It was the tremor not of a nervous man, but of a finely tuned, high-performance competitor in the war of ideas: Darwin's greyhound."[36]

Dawkins in his new role vilified those who in his view challenged science and rationality. In *River out of Eden* (1995), he had already attacked the extreme postmodernist claim that, in his words, "science has no more claim to truth than tribal myth." His response was: "Show me a cultural relativist at thirty thousand feet and I'll show you a hypocrite. Airplanes built according to scientific principles work."[37]

He continued to criticize cultural relativists and postmodernists as enemies of science. They were purveyors of "the low-grade intellectual poodling of pseudo-philosophical poseurs [that] seems unworthy of adult attention."[38] These writers all obeyed what he called Dawkins's Law of the Conservation of Difficulty: "obscurantism in an academic subject expands to fill the vacuum of its intrinsic simplicity."[39] In *Unweaving the Rainbow*, first published in 1998, he attacked writers who wrote antiscience screeds because they were "personally anguished, almost threatened, beleaguered, fearful of humiliation because science is seen as too difficult to master."[40]

The eminent science writer Timothy Ferris in a *New York Times* review of the book—headlined "Frauds! Fakes! Phonies!"—placed the book in a culture where, at the same time, popular antiscientific television shows investigated paranormal activities and elite books like *Higher Superstition* attacked postmodernist writings on science. But Ferris argued Dawkins's book would have limited impact: it would "be read mostly by college professors who already agree with [him]."[41]

But *Unweaving the Rainbow* did more than attack. It articulated in detail what had become the central rhetorical aim of his work in the mid-1990s: he wanted to replace a misguided, malignant faith in anti-rationalist mysticism with a rational appreciation of the glory of the natural world. In the book he aimed to create "poetic science." That is, a type of writing where scientists and poets challenged irrational delusions about how the world worked and produced their own descriptions of the beauty and majesty of all natural phenomena, from relativity to rainbows. Science was poetic. Science evoked wonder. Science revealed beauty.

Journalists divulged details of his personal life. Emblematic was a long interview in the *Guardian*. It described the large house in North Oxford he shared with Lalla Ward, which was accessible "by one of two gaps in a wall, scrunching over gravel through which bits of grass grow tastefully but not too tidily around the edges." Dawkins was described as "one of those fortunate men in whom, despite catkin-white eyebrows and the graying hair of a 56-year-old, you can still see the face of his boyhood." The reporter asked him how his life and writing changed after he had a daughter. "I don't see that much of her, to my enormous regret. I only see her alternate weekends," Dawkins responded. "You're so busy trying to make sure the weekend is a success, and that things don't go badly wrong, you don't have the luxury of exploring those other things."

The piece identified what it called "the Dawkins paradox." He wrote with a forceful directness, yet the reporter found him unengaging in person. The reason? "Probably," wrote the journalist, "it's the combination of that maddening Oxford air of high intellectual superiority . . . attached to an acute personal sensitivity." But writing about the beauty of science presented difficulties for Dawkins. Already his public image, in the words of the journalist, carried "so much baggage" that he could not write about the wonders of nature "without resuming the fierce diatribes against religion or sardonic attacks on other evolutionists whom he regards as misguided, which in great measure now define his public persona."[42]

This public persona was the subject of occasionally bizarre stories. After Dawkins said he would, in the right circumstances, have his daughter cloned, the *Daily Mail* asked if Dawkins was "the most dangerous man in Britain today." It called him a "militant atheist" and said "Dawkins, a handsome, plausible and self-confident performer on TV and radio, uses his position not only to undermine belief in God but to press the case for scientific adventurism of the kind many find frightening."[43] (Dawkins would later use a photo of the article as his online avatar.)

Other more sensitive writers and journalists contrasted Dawkins's public and private selves. Author Marek Kohn profiled Dawkins in his book on natural selection and Englishness. He described Dawkins as shy, with an "assiduously courteous" manner that creates "a line of defence which is conciliatory rather than antagonistic." He also noted how Dawkins's "widely noted good looks [were] set off to best effect by grooming and trimmings chosen with a stylish eye."[44] The *Guardian* said Dawkins's home looks more like a "playful library" where there are walls of books, "masks, model rabbits, birds' nests and wooden horses rescued from an ancient carousel," with a downstairs bathroom featuring framed doctorates, awards, and a decoration Dawkins downloaded from the Internet: a certification from the Universal Life Church.[45]

Prominent writers and journalists offered insights into his personality. Author and skeptic Michael Shermer said Dawkins was "somewhat shy and quiet," despite "his reputation as a tough-minded egotist."[46] British journalist Bryan Appleyard said Dawkins was "one of the strangest men I've ever known. . . . He is a highly strung, frequently petulant man. I've seen him storm out of an amiable dinner because he didn't like the music."[47] For journalist Ruth Gledhill, the scientist "in the flesh bears no resemblance to the angry, hate-filled anti-religionist he is portrayed as." To see him in person, she wrote, was "faintly transcendent."[48]

Appleyard offered a reason for his influence and popularity. "Dawkins is the supreme meta-establishment thinker," he wrote, "the eloquent defender of the dominant but seldom expressed worldview of our time—aggressive atheism and secularity, soft leftism, scientism and faith in progress. To his fans, he is reason incarnate."[49]

Part of his appeal, for other writers, was his combative intellectual style. Science journalist Robin McKie called Dawkins "the Dirty Harry of science"—"You don't assign him. You just let him loose."[50] The British political magazine *Spectator* gave his all-out attacks their own verb. "To be Dawkinsed," it wrote in a review of his 2003 essay collection *A Devil's Chaplain*, "is not just to be dressed down or duffed up: it is to be squelched, pulverised, annihilated, rendered into suitably primordial paste." But the magazine said Dawkins's attacks on those whom he deemed the enemies of truth were often too boorish and belligerent. "Dawkins," it wrote, "comes across as science's hired muscle: the bruiser in the bad suit with the baseball bat, stepping forward to administer a messy and unnecessary" death.[51]

Yet despite his academic title, Dawkins did not embrace the UK's burgeoning public understanding of science (PUS) movement. [52] In *Unweaving the Rainbow* he dismissed many PUS activities as the undignified dumbing-down of scientific work. "Funny hats and larky voices proclaim that science is fun, fun, fun. Whacky 'personalities' perform explosions and funky tricks," he wrote—activities that "betray an anxiety among scientists to be loved." A more realistic approach is to present science as a difficult subject that dedicated people must engage in a worthwhile struggle to master. As an analogy, the army "rightly don't promise a picnic: they seek young people dedicated enough to stand the pace." Dawkins said he did not condemn all science literacy activities—"only the kind of populist whoring that defiles the wonder of science."[53]

By contrast, Dawkins placed himself in a different tradition. "I'm of the Carl Sagan school of science writing," he would later say, "it should be beautiful and inspiring and enthralling and thrilling."[54] But his view of communication is not all about aesthetics. He argued also that science journalism was too important to be left to science journalists and that objective truth, discovered by scientists, deserved legal rights similar to libel laws to allow science to protect its reputation. He codified his communication philosophy in a memo he sent to Prime Minister Tony Blair in 2000. "If I am asked for a single phrase to characterize my role as Professor of Public Understanding of Science," he wrote, "I think I would choose Advocate for Disinterested Truth."[55]

THE FIRST CELEBRITY ATHEIST

Dawkins had written about religion as far back as *The Selfish Gene*. That book introduced his concept of memes, units of cultural reproduction analogous to genes, such as songs, religious beliefs, and folklore, to explain how cultural concepts replicated and survived as they were passed from generation to generation. Dawkins's main example of a meme was religious belief.

As his public career progressed, he critiqued religion more and more intensely and advocated for atheism. But 9/11 became a defining moment in his public career. The attacks that left three thousand dead after two planes flew into the World Trade Center led to a new intensity in his writing. He rejected the notion that people's respect for religion somehow sealed the topic off from criticism.

Four days after the attacks, he wrote in the *Guardian* that faith in the afterlife motivated the al-Qaeda hijackers. "To fill a world with religion, or religions of the Abrahamic kind, is like littering the streets with loaded guns," he wrote. "Do not be surprised if they are used."[56] He wanted to expand this idea immediately into a book, but an agent argued that the United States would not accept a book so critical of religion so soon after the attacks. Dawkins waited.

The God Delusion, first published in 2006, was an all-out attack on religion and believers. Dawkins argued that no rational basis existed for believing in a supernatural God and no evidence supported religious belief. But this was not just a personal matter: he argued religious belief halted progressive public life. He issued in the book a call to arms for atheism, seeking to convert believers and agnostics to atheism, listing in an appendix atheist support groups. The God Delusion was released in conjunction with Root of All Evil?, a documentary first broadcast on Britain's Channel 4 that presented religion as a malignant force in the world. Asked about the intensification of this strand of his intellectual work, he told the Guardian: "I seem—I seem to have lost patience."[57]

The book polarized reviewers. Scholars criticized what they considered his weak understanding of religion. Philosopher Thomas Nagel in the New Republic dismissed Dawkins's philosophical work as amateurish.[58] Evolutionary biologist H. Allen Orr in the New York Review of Books noted that the detailed exposition and attention to nuance in Dawkins's early writing on science had disappeared and, referring to religion, he was "a blunt instrument, one that has a hard time distinguishing Unitarians from abortion clinic bombers."[59] For Orr, The God Delusion was not a work of science or evolutionary biology. "None of Dawkins's loud pronouncements on God follows from any experiment or piece of data," he argued. "It's just Dawkins talking." In the London Review of Books, cultural critic Terry Eagleton noted Dawkins had no theological training. Moreover, he argued the book might have been better had its author avoided being "the second most frequently mentioned individual in his book—if you count God as an individual."[60]

Dawkins's supporters defended the book. Daniel Dennett said the book was not intended as a contribution to philosophical theory, but was rather a consciousness-raiser for the general reader.[61] Nobel-winning physicist Steven Weinberg said it was unfair to attack Dawkins for his lack of philosophical or theological training, as it was unreasonable to expect only experts to comment on public affairs.[62] Novelist Ian McEwan applauded Dawkins's unabashed intellectual honesty.[63] In a preface to the paperback edition, Dawkins castigated what he called the weak liberalism of atheist reviewers who believed in belief. He argued that the book's targets—the evangelicals

Jerry Falwell and Pat Robertson, as well as Osama bin Laden and the Aya-tollah Khomeini—were not unrepresentative caricatures, but influential and persuasive presences in cultural life.[64]

THE FIGUREHEAD FOR A
NEW SOCIAL MOVEMENT

The God Delusion connected with citizens, selling more than one and a half million copies worldwide. It became an iconic text for atheists worldwide who lacked a public spokesman, despite the fact that fewer than one in ten people in Western Europe attend church and one in seven U.S. citizens consider themselves utterly indifferent to religion.[65] Even so, atheists have had a limited space in U.S. public life and also attracted significant personal and social stigma.[66] The book's success, for Michael Shermer, showed a "market testimony to the hunger many people—far more, I now think, than polls reveal—have for someone in a position of prestige and power to speak for them in such an eloquent voice."[67]

The book helped spark a social movement. Variations of Dawkins's ideas appeared in books published around the same time by prominent intellectuals: *The End of Faith* (2004) by Sam Harris, *Breaking the Spell* (2006) by Daniel Dennett, and *God Is Not Great* (2007) by Christopher Hitchens. The four books together constituted the foundational texts of a new social movement—new atheism.

New atheists attacked religion directly, aimed to persuade nonbeliev-ers to declare themselves atheists, argued that faith was unjustified belief unsupported by evidence, and believed religion was doomed. They argued that science was the only means of knowing, the way to cure faith and lead people to rational enlightenment. The movement aligned itself with the long-standing work of skeptics and secular humanists who advocated in public for an understanding of the world based on reason, evidence, and common sense, not supernatural forces.[68]

These writers saw themselves as agents of social change. Similar to the efforts of the feminist and gay rights movements, they aimed to mobilize marginalized citizens, trying to persuade nonbelievers to come out as athe-ists.[69] The four new atheist books, scholars of religion argued, became cul-tural touchstones that brought together a diverse global group of secularists who used the texts to define and deepen their atheism.[70]

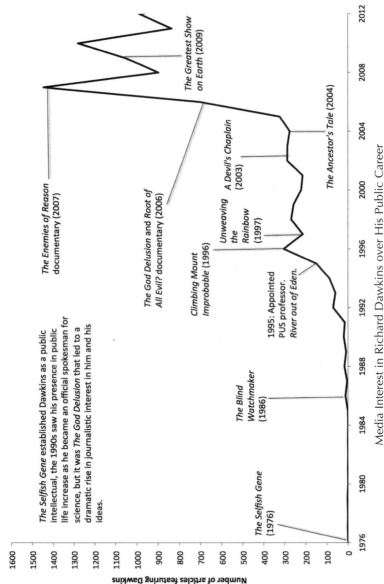

Media Interest in Richard Dawkins over His Public Career

Dawkins became the figurehead for this new social movement. *Wired* called him new atheism's "leading light."[71] With *The God Delusion*, as the chart shows, worldwide media interest in Dawkins soared.[72]

The popularity of his book and his enormous media exposure allowed Dawkins to build a community around his advocacy of science atheism. In 2006, he founded the Richard Dawkins Foundation for Reason and Science. When it was first established, Dawkins spelled out its mission: "The enlightenment is under threat. So is reason. So is truth. So is science, especially in the schools of America," he wrote. "I am one of those scientists who feels that it is no longer enough just to get on and do science. We have to devote a significant proportion of our time and resources to defending it from deliberate attack from organized ignorance."

He also established Richarddawkins.net, a place where his supporters could gather to discuss atheism and read about all things Dawkins-related. Fans could buy books, DVDs, audiobooks, tote bags, mugs, bumper stickers, *God Delusion* T-shirts, hooded tops, and—to identify themselves as atheists—lapel pins in the shape of an *A*. They could also read descriptions of the Out campaign that urged nonreligious believers to declare their atheism publicly. They could print fliers, meet other atheists, and leave messages for Dawkins—which he regularly answered.

Supporters online expressed their admiration and adulation for Dawkins. Among the comments viewed on July 3, 2009, on the official YouTube channel of Dawkins's site, were: "Mr Dawkins you are amazing. I wish more people would see your point of view and maybe then we could all make a better world" (from bloodlover); "Long live Richard Dawkins. I love Dawkins to bits. . . . Thank you for fighting the good fight sir. Thank you for bringing Atheism out in the open and making it more accessible to the common man" (from reytrue); "Dawkins himself is a God. God of the Athiests" [*sic*] (spartanses); "If I ever shake hands with Dawkie, I assure you, I will NEVER wash it again! I'd have genius DNA on my hand! To me, it would be like meeting Plato or Einstein." (from silviafarfallina).[73]

Dawkins, for one sociologist of religion, came to occupy a symbolic role similar to evangelists. Dawkins was a role model for atheists. *The God Delusion* not only transmitted ideas, but was a book around which nonbelievers built their identities as atheists. Dawkins's media profile and commercial products helped validate atheism as a way of life.[74]

A publishing sub-industry formed around *The God Delusion*. Among the counterarguments was writer John Cornwell's *Darwin's Angel: An*

Angelic Riposte to "The God Delusion" (2007) and Oxford theologian Alister McGrath's co-authored *The Dawkins Delusion? Atheist Fundamentalism and the Denial of the Divine* (2007). Among the anti-Dawkins books have been: *God Is No Delusion: A Refutation of Richard Dawkins* (2007) and *Challenging Richard Dawkins: Why Richard Dawkins Is Wrong about God* (2007).

At the same time, Dawkins communicated these ideas also through traditional media. He hosted the two-part documentary *The Enemies of Reason* (2007), where he debunked various forms of pseudo-science. He was a major figure in the story; viewers saw him have his aura photographed, have a tarot card reading, and have his chakras energized by illuminated crystals. *The Genius of Charles Darwin* (2008), a three-part documentary, described the naturalist's life and theories. In 2006, a celebratory book of essays was published: *Richard Dawkins: How a Scientist Changed the Way We Think*, a way of reinforcing his scientific status and influential reputation.

Dawkins's thinking about science communication remained unchanged. For example, he appeared at a 2006 conference on science and religion, an event science writer George Johnson described as less of a polite dialogue than "the founding convention for a political party built on a single plank: in a world dangerously charged with ideology, science needs to take on an evangelical role, vying with religion as teller of the greatest story ever told."[75]

At the conference, Neil deGrasse Tyson argued that Dawkins's combative communication style reduced his potential to persuade even greater numbers of people about the value of science. "You're professor of the public understanding of science, not professor of delivering truth to the public, and these are two different exercises," Tyson told Dawkins. "Persuasion isn't always 'here's the facts, you're either an idiot or you're not.' It's 'here are the facts and here is a sensitivity to your state of mind,' and it's the facts plus the sensitivity, when convolved together, creates impact."

In response—to hoots of audience laugher—Dawkins said: "One anecdote to show that I'm not the worst in this thing: A former and highly successful editor of *New Scientist* magazine . . . said, 'Our philosophy at *New Scientist* is this: science is interesting, and if you don't agree, you can f**k off.'"[76]

His ideas spread through popular culture via unusual roots. He appeared, as himself, in a 2008 episode of *Doctor Who*, where his first words in a television interview were: "It's an empirical fact." He was caricatured in the satirical cartoon *South Park* in the episode "Go God Go," where he was the high school's new teacher of evolution.

"So you *are* saying that we're all related to monkeys," he is asked in class.

"Well, yes, basically we are."[77]

THE SELFISH GENIUS?

Dawkins's rise to fame was not without major criticisms. His self-crafted role as public spokesman for Darwin was challenged. *The Selfish Genius* by historian of science Fern Elsdon-Baker argued Dawkins "hijacked" and "distorted" Darwin's legacy by presenting the gene-centered view as central to contemporary evolutionary theory when the discipline has not completely conformed to this perspective. There have been challenges to the neo-Darwinian orthodoxy, especially through the study of the genetics of bacteria and microorganisms, where horizontal gene transfer, the movement of genes between nonrelated organisms, has made more complex the view of genes flowing down through generations. It was accepted within evolutionary biology, wrote Elsdon-Baker, that he does not speak for the field. Some evolutionists disputed whether he was even a scientist.[78]

As his public career progressed, his original scientific ideas have only exerted a limited influence among biologists. He viewed the final four chapters of *The Extended Phenotype* as his novel contribution to evolutionary theory, but the theory has not developed into a research theme in any scientific institution. Dawkins said he has not been able to expand its research agenda because he has been preoccupied with writing books.[79] But there are signs of renewed interest. A 2008 European Science Foundation workshop brought scientists together to have a fresh look at the extended phenotype. A press release about the conference said it "resurrected" the idea as "an important and valuable concept that helps explain evolution."[80]

His idea of the meme has had a complicated reception. Other researchers developed the idea, but it never gained traction within science. An online *Journal of Memetics* faded away after eight years of infighting over objectives and terminology. Memes, researchers argued, did not lead to a mass of research papers and failed to generate original observations or experimental data and were therefore a "scientific flop."[81]

But with the rise of Internet culture from the late 1990s, the concept flourished as a way to describe how online ideas spread and mutated. Internet memes are different from *The Selfish Gene*'s memes: they do not mutate randomly—they are manipulated deliberately by creative humans. *Meme* became a buzzword in the Internet era, wrote science writer James Gleick in his history of the concept of information. Memes were the perfect description of how information spreads now, in what Gleick called our "age of virality," where a few keystrokes move ideas instantly across the connected globe. For Gleick, the meme was Dawkins's "most memorable invention, far more influential than his selfish genes or his later proselytizing against religiosity."[82]

A part of the atheist community has distanced itself from Dawkins. When Daniel Trilling took over as editor of *New Humanist* magazine, the publication of the British Rationalist Association, he called for a new discussion of religion—one that moved beyond "personality clashes," or a winner-takes-all bout between science versus religion. A new approach is needed to negotiate ethical and political issues while upholding secular values. It is a difficult task. "Currently," wrote Trilling, "Richard Dawkins is providing a case study in how not to do it."[83] And in 2012, physicist Peter Higgs said Dawkins adopted a "fundamentalist" approach to dealing with believers, and he said science and religion could be compatible.[84]

After *The God Delusion*, Dawkins returned to writing about the science that made his name: evolution. He wrote to fill what he considered a gap in his oeuvre: a book that explained clearly how evolution occurred. *The Greatest Show on Earth* (2009)—published on the anniversary of Darwin's birth, and the 150th anniversary of *The Origin of Species*—became a *New York Times* best seller. Dawkins closed the book with a warning about the continued social importance of writing on evolution. At least 40 percent of Americans, he wrote, are "dyed-in-the wool, out-and-out, anti-evolution creationists," adding that in Britain the numbers are less extreme, but not encouraging. He wrote: "There are still no grounds for complacency."[85] He aimed his next book at children, to instill the same sense of wonder he felt at nature, an idea he encapsulated in the book's title, *The Magic of Reality*.

Meanwhile, his advocacy work continued. The Richard Dawkins Foundation for Reason and Science aims "to realize Richard's vision to remove the influence of religion in science education and public policy, and eliminate the stigma that surrounds atheism and non-belief."[86] In recent years also, Dawkins's online presence has expanded dramatically. He has more than one million Twitter followers. His Twitter self-description encapsulates his public career—and notably seeks to soften his attitude to religion: "English biologist & writer. Likes science, the poetry of reality. Dislikes pretentious obscurantism. Treats all religions with good-humoured ridicule." He continued to make documentaries with Channel 4. *Faith School Menace* examined the harmful effects that schools with a religious character have on students and society (2010). *Sex, Death and the Meaning of Life* (2012) explored how science and reason can address life's big issues.

Yet his public image over time crystallized around his atheism. The change in his public persona can be seen by tracing the changing labels journalists used to describe him. Over twenty years in the *Times*, for example, descriptions of Dawkins have moved from "The Zoology Man"[87] and "lecturer in zoology at Oxford"[88] to "a biologist whom people love to hate"[89] to "the first Oxford professor of the public understanding of science"[90] to "Darwinist proselytiser"[91] to "Britain's best-known Darwinist"[92] to "Britain's best-known atheist"[93] to "one of the great atheists of our time,"[94] "the famous atheist,"[95] "scientific atheism's heavyweight champion,"[96] "Britain's angriest atheist and self-appointed Devil's chaplain,"[97] "the media village's number one atheist,"[98] and "biologist and arch atheist."[99] When Dawkins met Stephen Hawking on-screen as they interviewed each other as part of a television show on the UK's scientific heritage, he was asked by the physicist: "Why are you so obsessed with God?"[100]

An Appetite for Wonder was a vehicle for Dawkins to reshape his public image. It allowed him to present the softer, more human side of his complex character that his friends say has been obscured in many of his public portrayals. Dawkins described his childhood, his school years at a British public school, and his years of training as an ethologist. The book was an intellectual biography, one that traced his scientific influences and academic development. It also reinforced his research credentials: its subtitle was *The Making of a Scientist*. Dawkins made few personal revelations, except for his account of losing his virginity, aged twenty-two, to a cellist. He wrote: "It isn't that kind of autobiography."[101]

Reviewers overall welcomed his thoughtful self-portrayal. "The Richard Dawkins that emerges here," the *Independent* wrote in a representative review, "is a far cry from the strident, abrasive caricature beloved of lazy journalists with an op-ed piece to file."[102] The *New York Times* identified the particular difficulty faced by Dawkins, who had written so prolifically for decades. The paper noted the book contained stories he told elsewhere, excerpts from quotations he gave elsewhere—it read "like the work of a man who has already written abundantly about himself."[103] But the book evoked hostility: for the *Sunday Times* it was an unenlightening type of "self-promotion without self-knowledge." To illustrate this point, the review focused on the book's conclusion where Dawkins compares his qualities with those of Darwin. "He says he does so 'in humility,'" the reviewer wrote, "but it's an irritating exercise in fake self-deprecation, like so much of his writing here."[104] Nevertheless, readers responded. The book entered the *New York Times* hardcover nonfiction best seller list at number 11— testament to the interest Dawkins continues to hold for general readers.

While Dawkins showed those readers parts of his private self, he continued his sharp-edged atheist advocacy. The documentary *The Unbelievers* followed Dawkins and cosmologist Lawrence Krauss on a tour to promote atheism (and sell books). It ended with the Reason Rally in Washington, D.C., where Dawkins looked out over the crowd and called it "the most incredible sight I can remember ever seeing."[105] The *Los Angeles Times* called the film a "high-minded love fest between two deeply committed intellectuals and the scads of atheists, secularists, free-thinkers, skeptics and activists who make up their rock star-like fan base."[106]

Yet his influence has waned. The public controversies within which he forged his reputation have faded. In 2013, the British current affairs magazine *Prospect* named him the world's most significant thinker. In 2014, it did not put Dawkins in the top 50.

THE CELEBRITY SCIENTIFIC INTELLECTUAL

Stephen Hawking embodied the life of the mind, but he has remained a working scientist whose contributions to public thought have been restricted to the simplified explanations of complex physics. As science writer Philip Ball noted, he has no reputation among scientists as a deep thinker on the myriad ways science impacts public life.

Richard Dawkins, however, is the archetypal celebrity scientific intellectual. He became an expert in ethology at the Oxford zoology department that was an acknowledged intellectual force in evolutionary thinking. With *The Selfish Gene*, he spread the new biological ideas of genetic evolution through wider society, topics that connected the book and its young author with incendiary ideas in the late 1970s about the unresolved debate over whether an individual's biology determines one's place in society.

He continued to feature as a protagonist in many of the major clashes of ideas in the modern era. As creationism rose as a political force, Dawkins became a stout public defender of evolution. As evolutionary ideas rose to be influential ways of understanding life and society, he became with Stephen Jay Gould the public face of Darwinism. As science in the 1990s believed itself under siege by the massed forces of pseudoscience, antiscience, and postmodernism, Dawkins became the hawkish emblem of a muscular science, a sharp-elbowed science advocate. As fears of religion's malignant influence rose post 9/11, he became the world's most famous—and most famously aggressive—atheist.

Each of these roles folded into the next, as they were different ways of expressing the need in public life for reason, rationality, and evidence, the foundational features of science. He formed a deserved reputation as a clear thinker and writer, who had a rare ability to state complex ideas in provocative, sharp, and lucid prose that meant his ideas pierced through the noise of a crowded public arena to reach readers. At the same time as this style appealed to audiences, it also created an image of science as strident, intolerant, and authoritative—features that add little to scientists' desire to be appreciated more by swaths of the broader population.

As he became a public intellectual, the mechanics of celebrity started to work. He not only articulated and stood for the gene's-eye view of evolution, the power of Darwinism, the potency of science, and the logic of atheism, but he also came to *embody* those views in public. He became, in celebrity culture, a shorthand for these complex issues in public life. He made these abstract ideas human. Darwin's greyhound. Mr. Public Science. God of the Godless.

But the fame of Dawkins also illustrates a conundrum for the celebrity scientific intellectual. To appear on the public radar, a star needs to create a clear image of himself or herself in the public mind. Once that image is established, it is almost impossible to lose that persona and create an entirely new one. That's why it was a seamless movement for Dawkins to go from "The Dirty Harry of Science" to "Britain's angriest atheist."

More problematic, however, was the attempt in his autobiography to present a more personable Dawkins. One image does not replace the other. They remain as opposites that jostle each other in the public arena. Dawkins's image as hard-headed defender of rationality becomes blurred when he presents a softer image. That's his image problem, an unfortunate side effect of the need for stars to have a clear signature in celebrity-driven public life.

FOUR

Steven Pinker's
Academic Stardom

The world can peer into Steven Pinker's genes. In 2008, the Harvard psychologist became one of the first ten volunteers to donate his DNA to his university's Personal Genome Project, the enterprise that wants to comb the genetic blueprints and medical histories of 10,000 people to predict how humans develop or avoid disease. Pinker, moreover, allowed researchers to post his genome on the Internet. He also published a 7,974-word magazine article that interpreted his raw genetic data for general readers. The man the *New York Times* once called "Nature's Pathologist" put his personal genome on the slab, and dissected it before an audience of millions.

Pinker—identified in the project as PGP6 and hu04FD18—revealed that he carried one copy of a gene for familial dysautonomia, an incurable disorder of the nervous system that has a high chance of causing premature death, a finding significant to his extended family. "Children are not in my cards," he wrote in the long *New York Times Magazine* article, "but my nieces and nephews, who have a 25 percent chance of being carriers, will know to get tested."

The genetic blueprint revealed the historical migrations of his distant ancestors. Montreal-born Pinker descended from Jewish exiles who fled to Italy after the Romans destroyed Jerusalem, then to the Rhine Valley, then to Poland and Moldova. "And even this secular, ecumenical Jew experienced a primitive tribal stirring in learning of a deep genealogy that coincides with the handing down of traditions I grew up with," he wrote. "But my blue eyes remind me not to get carried away with delusions about a Semitic essence."[1]

Pinker, however, refused to find out his risk of Alzheimer's. But he did note that he enjoyed beer, broccoli, and Brussels sprouts even though he carries genes for tasting their bitterness. And he found he had a gene that gives him typical odds for red hair even though his shoulder-length black-gray curls had long prompted journalists to compare him with long-haired 1970s rock musicians Robert Plant, Roger Daltrey, and Jimmy Page.

Personalizing the aspirations and anxieties of a new genetic society, Pinker's tale hit a cultural nerve. It was reprinted in *The Best American Essays 2010*, *The Best American Science Writing 2010*, and in *Ethical Issues in Modern Medicine* (2012). The article challenged the utopian promise of personalized medical treatments and the dystopian fears of a *Brave New World*–like society where a genetic underclass is corraled into menial work and denied life or medical insurance.

The insightful, provocative piece typified the writings that made Pinker one of the world's leading public intellectuals. He has been hailed as the foremost explainer of linguistics to popular audiences, lauded as an influential scholar who has made lasting contributions to the fields of language development and visual cognition, praised as an advocate for Darwinism, and recognized as a persuasive ambassador for the intellectual power of science.

He has also been called one of "language's bad boys," "the world's foremost rock 'n' roll neuroscientist," and the "stud-muffin of science." Unlike Richard Dawkins, he remained in the lab, and unlike Stephen Hawking, his popular books were not just explanations of ideas, but controversial interventions into public affairs. "Pinker is a star," wrote Dawkins, "and the world of science is lucky to have him."[2] Controversial and celebrated, popular and prolific, Pinker is the exemplar of the modern superstar academic.

RESEARCHING AT THE DAWN OF COGNITIVE SCIENCE

The study of the human mind was transformed in the late 1950s. Until then behaviorism dominated twentieth-century psychology. Behaviorism—associated chiefly with Harvard psychologist B. F. Skinner—explained actions as the conditioned responses to external stimuli. Rats, for example, learned to navigate mazes to find cheese. Dogs drooled when they heard a ringing bell they learned to associate with food. The field changed dramatically after psychologists flooded their work with new ideas from the embryonic fields of computer science, information theory, artificial intelligence, and centuries-old philosophies of mind. Researchers coalesced around fresh, shared ideas that knowledge was a form of information and thinking was a type of computer-like information processing. Born was the cognitive revolution. Born was the new field of cognitive science.

The new science revolutionized linguistics. Until the late 1950s, linguistics operated under the consensus that culture created language. Children, specialists agreed, acquired language from adults and other children. Linguists learned how language worked by studying different spoken languages to find patterns of words, sounds, and sentences. From these patterns they produced dictionaries and grammars.

But one young linguist disagreed and set out to demolish this consensus: Noam Chomsky. He argued that language was rooted not in culture, but in biology. All the rules for language—for all the world's tongues—were ingrained from birth in a newborn's brain. Babies were language machines, hardwired for grammar.[3]

Cognitive science and behaviorism were embedded in wider political ideas. The behaviorist approach was seen as counter to eugenics and the Nazi-claimed scientific justifications for biological Aryan superiority and Jewish inferiority.[4] And some of the motivating ideas of the civil rights movement, furthermore, meant scientists were cautious about claims that humans had an innate nature. A legacy of the 1960s, Pinker later wrote, was an antibiological view of human nature, resulting from one of the era's leading ideas that reforming social ills meant reforming social institutions.[5]

Pinker studied psychology after the cognitive and Chomskyan revolutions. "At the time the cognitive psychologists were considered upstarts and revolutionaries," he would later recollect, "and it was fun to get into what at the time was an insurrection."[6] He did his undergraduate work at McGill University, and then graduated from Harvard in 1979 with a PhD in experimental psychology. Visual cognition was his doctoral specialization; he worked on how the brain processes information received by the eye, in particular how three-dimensional space is represented as mental images. But it was at Harvard that he also first studied a subfield of psychology that would shape his life's work. It was an area filled with contesting theories, one of the most fraught and difficult areas of developmental psychology: how humans acquire language.

But he found the theories of how children learned language "vague and squishy," unlike the clean, definite, and mechanistic models that he was used to from his studies on vision and memory. Pinker wanted to put theories of baby-talk on a similarly firm scientific footing. Applying concepts from mathematics and computer science, his first single-authored paper, published in 1979, drew on Chomsky's ideas to identify precise mechanisms through which language is learned.[7]

Pinker quickly established himself as a rising star in cognitive science. In 1982, after stints at MIT, Harvard, and Stanford, he settled down to a position in MIT. In 1984, the American Psychological Association gave him a prestigious award for early career contribution to the field, and Harvard University Press published his first book, a work on children's language development, *Language Learnability and Language Development*. Pinker would go on to author another book on language, and edit or coedit three other specialist books in the areas of language and visual cognition.[8] MIT tenured him in 1985. Publications outside the academy took notice: in 1986, *Esquire* labeled him one of its outstanding men and women under forty.

LANGUAGE'S BAD BOY: BRINGING LINGUISTICS TO SOCIETY

The linguistic revolution failed to grab the public imagination. Chomsky is today best known in public life for his trenchant left-wing criticism of American media and foreign policy—not as the towering figure in modern linguistics. His peers blamed his prose style that is so dense that even professional linguists often found it difficult to read. As a consequence, linguistics was trapped in a paradox: its discoveries were central to the human condition, yet were communicated in language understandable only to professional linguists. The field needed clear communication.[9] The field needed a popularizer.

Pinker looked perfect for the role. After he published his second academic book, he later wrote, "an editor told me (I am paraphrasing) that my writing did not suck and encouraged me to reach a wider audience."[10] He started writing a popular book about language one summer and for the only time in his career kept to a furious pace of one chapter per week.[11]

The Language Instinct, first published in 1994, brought Chomsky's ideas to a wider audience. He conveyed the core Chomskyan argument that the roots of language were biological. People knew how to speak, wrote

Pinker, "in more or less the sense that spiders know how to spin webs."[12] Pinker integrated ideas from linguistics, cognitive neuroscience, developmental psychology, and speech therapy to challenge other conventions and preconceptions about language. He argued that language did not shape thought, but instead people—whether they speak English or French or Apache—shared a common universal language of thought that Pinker called mentalese, which existed prior to speech. He made linguistics vibrant and interesting to nonreaders: "The word *glamour* comes from the word *grammar*," he wrote, "and since the Chomskyan revolution the etymology has been fitting."[13]

With the book, Pinker carved out his own intellectual space. Chomsky argued language was an unavoidable by-product of the complex brain's evolution. By contrast, Pinker argued language evolved an adaptive trait: it was shaped by evolution to achieve an advantage in passing genes to the next generation. The hunter-gatherer ancestors who best used language had an advantage over others when it came to survival and reproduction. *Nature* would later summarize the approach as "Survival of the Clearest."[14]

The book was a landmark in the public understanding of language. One linguist wrote in *Nature* that it was "one of the best things to have happened in the dissemination of ideas in linguistics and cognitive science."[15] Another linguist said the book established Pinker as "the preeminent interpreter of our discipline for general audiences."[16] But another linguist warned readers that Pinker presented a particular view of a vast field. "He is lobbying for Chomsky's theory, not describing the entire field, nor reporting a consensus," wrote Randy Harris in the *Globe and Mail*. "Just remember he is not only explaining. He is also selling, which means he is . . . making it a little sexier than it really is."[17] Nevertheless, the *New York Times* named it a 1994 editor's book choice, the Linguistic Society of America gave the work its public interest award, and *American Scientist* named it as one of its top 100 science books of the twentieth century.

The book—published in hardcover of 35,000 copies, and excerpted in the *New Republic*—displayed Pinker's qualities as a thinker and writer. He characterized himself as "an opinionated, obsessional researcher who dislikes insipid compromises that fuzz up the issues," who writes "with a passion for powerful, explanatory ideas, and a torrent of relevant detail."[18] And reviewers found he had similar skills to other famous popularizers. "In a manner reminiscent of the work of the paleontologist Stephen Jay Gould," wrote the *New York Review of Books*, "Pinker moves back and forth between

The *Boston Globe* photographed Steven Pinker in 1994, around the publication of his first popular book *The Language Instinct*, with one of his personal collection of tabloid stories that reflect his argument that humans are hardwired from birth to use language. Boston Globe via Getty Images

the questions and concerns of the intelligent lay reader and those of the specialist."[19] Popular science writer John Gribbin wrote in the *Sunday Times*: "He does for language what David Attenborough does for animals, explaining difficult scientific concepts so easily that they are indeed absorbed as a transparent stream of words."[20] Pinker later identified the book as "a turning point in my professional life."[21]

The Language Instinct jump-started Pinker's celebrification. The *Boston Globe* called him one of "Language's Bad Boys" in an article that it illustrated with a picture of Pinker reading a copy of the British *Sun* from his collection of tabloid coverage of his research topics. The journalist linked Pinker's idiosyncratic appearance to his provocative writing. "With his shoulder-length frizzy gray-black hair, MIT professor Steven Pinker, 39, looks a bit like a young Albert Einstein without a moustache," the piece began. "Otherwise he resembles a professor less than a smart and sassy grad student in jeans. That somewhat off-the-model style fits in well with the book Pinker has just written, a profound, puckish and iconoclastic study of how and why we use language."[22]

Looking back, Pinker said this characterization surprised him a little. He said: "I knew that journalists had to have a hook . . . something distinctive that makes for livelier journalism. . . . I was an avid enough consumer of journalism to know that journalists like to dress up a story."[23]

The *Montreal Gazette* focused similarly on his personal appearance, noting that his "shock of curly hair falls to his shoulders." It also described his domestic life, noting he lived alone in an upstairs apartment on a quiet street, but reported that the "vase of fresh tulips, resting on the small dining table near a painting of irises, suggests his fondness for flowers." The *Gazette* made a bold prediction about Pinker's future. "Having long held the respect of his peers," the paper noted, "Pinker now stands poised to become one of the best-known scientists in the United States."[24]

AMBASSADOR FOR EVOLUTIONARY BIOLOGY

The Language Instinct revealed Pinker as a fearless intellectual fighter. In the book, he took on virtually all social scientists for what he saw as their enforcement of a suffocating intellectual dogma—the denial of human nature. This disavowal formed the cornerstone of what two scientists, the wife and husband team Leda Cosmides and John Tooby, called the Standard Social Science Model (SSSM), a contemporary academic orthodoxy that held that humans were shaped almost exclusively by their education and experience, a viewpoint that for Pinker had become "the secular ideology of our age, the position on human nature that any decent person should hold."[25]

But Pinker—following Cosmides and Tooby—said there was an alternative way of seeing the world, one that explained how evolution forged the brain and caused humans to learn and acquire values, an alternative view that integrated psychology, anthropology, neuroscience, and evolutionary biology into a new way of knowing that embraced the biological reality of human nature. Cosmides and Tooby, leaders in one faction of this new science, referred to it as the Integrated Causal Model. They also called it evolutionary psychology.

Evo-psych quickly became an on-the-pulse intellectual trend of the 1990s. Its core tenet was that evolution designed the human mind to survive on the African savannah in the Pleistocene era, between ten thousand and 1.7 million years ago, as small bands of hunter-gatherers sought food, safety, and sex. The mind was a Swiss Army knife, in Cosmides and Tooby's metaphor, filled with various tools for different tasks that natural selection designed to solve problems in the ancient landscape in which it evolved. The modern skull, essentially, houses a Stone Age mind. The *New Yorker* later labeled this much-repeated phrase the field's "snappy slogan."[26]

Biologists more generally are cautious about identifying particular traits as adaptations, because evolution is so complex. But based on the tenets of evo-psych, all sorts of researchers—psychologists, anthropologists, economists, historians, literary theorists, and other scholars—came together under the framework of evolutionary psychology, seeking to understand human thought and behaviors in Darwinian terms. *Time* magazine reported on the field and its ideas in a cover story.[27] By the end of the decade *The Language Instinct* was one of several books—others included Terrence Deacon's *The Symbolic Species*, Michael Gazzaniga's *Nature's Mind*, and Daniel Dennett's *Darwin's Dangerous Idea*—published on evolutionary psychology for the general reader.

While on sabbatical at the University of California with Cosmides and Tooby, Pinker wrote *How the Mind Works* (1997). The 660-page book expanded on a theme of *The Language Instinct*, positing that if language is an evolved instinct, then other aspects of the mind are too. Pinker argued that visual cognition, thinking, art, music, literature, politics, and friendship can be revealed through insights from evolutionary psychology. Art grants us status. Friends help us out when we need it (in return for our future re-ciprocal aid). Young women seek rich men who can care for their offspring. Men seek nubile young women who will bear them healthy children.[28] Evo-lutionary psychology, he wrote, could explain religion as a way to enhance people's belief in survival.

In the book, Pinker drew on personal experience to show how humans can override their biological drives. "Well into my procreating years I am, so far, voluntarily childless, having squandered my biological resources reading and writing, doing research, helping out friends and students, and jogging in circles, ignoring the solemn imperative to spread my genes," he wrote. "By Darwinian standards I am a horrible mistake, a pathetic loser. . . . But I am happy to be that way, and if my genes don't like it, they can go jump in the lake."[29]

With the book, Pinker became the most prominent, most pugnacious, and most eloquent spokesman for the new field. The science journalist John Horgan wrote that perhaps Cosmides and Tooby's greatest feat was draw-ing Pinker into the field, as his background as a cognitive scientist granted the field scientific rigor, or the impression of rigor, and his skill as a writer ensured he could communicate clearly its central ideas.[30] Science writer Steven Johnson noted that Pinker had "made a name for himself as one of evolutionary psychology's most appealing ambassadors."[31]

Influential reviewers lavished praise on *How the Mind Works*, which became a finalist for the Pulitzer Prize in nonfiction. Christopher Lehmann-Haupt in the *New York Times* concluded: "As lengthy as it is, it will produce a book in the reader's head that is even longer. For it alters completely the way one thinks about thinking."[32] Writer Marek Kohn predicted that its synthesis of cognitive science and Darwinism would make it "a landmark in popular science."[33] Science writer Oliver Morton in the *New Yorker* said it "marks out the territory on which the coming century's debate about human nature will be held."[34] Michael Gazzaniga, a leader and pioneer in cognitive science, wrote that the book had implications for how scientists in the field do their work. "Upon full consideration of its message," he wrote in *Trends in Cognitive Sciences*, "cognitive scientists can examine what it is they are trying to do with greater clarity and neuroscientists can begin to ask a different kind of question." The book, he noted, pointed to the future of scientific work. "The jig is up," wrote Gazzaniga. "Wallowing around in one's subdiscipline won't work anymore."[35]

However, other scientists criticized the book, their arguments reflecting the general objections to evolutionary psychology as a whole. Geneticist Steve Jones in the *New York Review of Books*, for example, said Pinker put too much emphasis on the biological roots of behavior. Pinker, he said, had an "inclination to overbiologize the human race." Psychologists also risked putting too much faith in a compelling new explanation for what has always been messy human behavior. "In its early days [psychology] was intrigued by the idea—ludicrous in retrospect—that human society arose from the unconscious desire of sons to sleep with their mothers," he wrote. "Now there is a more subtle temptation; that the mind works the way it does because their great-grandmothers gathered berries."[36]

The book stepped-up Pinker's celebrification. The *Irish Times* opened its interview—headlined "Evolutionary Rock Star"—by asking why he did not conform to the idea of an MIT professor. "Maybe it's the scuffed cowboy boots. Or the shoulder-length curls," the reporter noted. "The diffident 43-year-old seems too young, too funny, too downright nice to have acquired that title."[37] The paper noted that Pinker, with his clear and exuberant writing, is the "ideal popularizer" of the brain, a trend in science publishing. It noted too that he had become "a celebrity scientist, the Carl Sagan of the human brain." Yet the journalist found that he was a committed teacher and researcher, far more than merely "a poster boy for cognitive studies."

Other journalists fawned over his appearance. A *Mail on Sunday* review noted that his "hairstyle has yet to evolve beyond 1975" and the "rugged, long-haired photo on the dust jacket suggested he might be the world's foremost rock 'n' roll neuroscientist."[38] In the *Times*, Elaine Showalter called him "a world-class cognitive psychologist and stud-muffin of science, with a winning smile, dimples and long curly hair."[39]

At the same time, other journalists picked apart Pinker's crafted public image. Science journalist John Horgan found Pinker "a bit too self-consciously packaged for a serious scientist. When I interview him, he is wearing black cowboy boots, black jeans, and a shirt whose turquoise stripes suspiciously match his eyes."[40] The *Times Higher Education Supplement* wrote: "The 43-year-old son of a travelling salesman, Steven Pinker's image is that of a populist academic with a flair for self-promotion."

Pinker himself responded to the frequent comments on his image. "My lifestyle is conservative. I'm not terribly different from other academics. I don't drive a motorcycle. I don't take drugs," he said. "Most of my life is spent writing, reading and going to conferences. I do have long hair, I wear colorful ties, I listen to rock music."[41]

BALANCING PUBLIC INTELLECTUAL WORK

Pinker, by the late 1990s, was an established scientific celebrity. The *Times* in 1999 called him "something of a star," noting that his books "made his youthful looks and rock-star haircut familiar on lecture platforms throughout the world."[42] He made a cameo as an MIT linguist in the novel *Infinite Jest* (1996), one of the decade's defining postmodern novels. A linguist wrote in the *New York Times* in 1999, "Pinker is now an established public intellectual, the most visible representative of the new field of cognitive science, a media star throughout the English-speaking world."[43] The scientist George C. Williams wrote: "I'm very favorably impressed with Steven Pinker. He's going to be a superstar well into the twenty-first century."[44]

With his next book, Pinker went back to his academic heartland. Conforming to the pattern of behavior of the public intellectual, after writing a controversial popular book, he shored up his research reputation by publishing a technical work. *Words and Rules*, first published in 1999, popularized his dozen years of research on the past tense of verbs. Learning a language, it argued, meant learning words and rules. When children start to speak English, they learn a rule: add -ed to the end of verbs to create the past tense. But when children apply this rule to one of the approximately 180 irregular verbs, they make mistakes: they say, in Pinker's example, "we holded the baby rabbits." So children must learn a new word—held—to speak correctly. Children soon become skilled in regular and irregular verbs because they have acquired words and rules, the basis of all language. Reviewers praised the book but found it considerably more technical than his other works. A cognitive neuroscientist, for example, in a *Nature* review called it "a reader-friendly balance between humour, irreverence and anally retentive scholarship."[45]

Despite being a technical work, *Words and Rules* still led to personality-focused media coverage. The *Guardian* sat in on his Psychology 101 class at MIT. "As the students gossip, the slight and compact figure of Steven Pinker arrives at the dais to start the lecture," the journalist wrote. "He is often described as looking like a rock star, and his curly shoulder-length mane and Cuban heels give him the air of a prog rocker on his third comeback tour. He has a superbly defined jaw, glittering blue eyes and a kilowatt smile which he beams at his class as he switches on the microphone." It described his marital history and noted: "Pinker is a private man . . . and when discussion shifts away from his work and ideas to his family, Pinker becomes quite withdrawn."[46]

At this time, too, he came to represent science more broadly. He was one of seven charismatic researchers profiled in *Me & Isaac Newton* (2000), a documentary film by acclaimed director Michael Apted about the nature and meaning of scientific work. Pinker was filmed conducting an experiment in a laboratory where he and a colleague identified the parts of the brain that lit up as a person spoke. For Pinker, being a scientist meant being controversial and provocative, "sticking your neck out" at risk of being wildly wrong.

After *Words and Rules*, Pinker fulfilled another obligation of the public intellectual: he wrote about what it meant to be a public intellectual. It meant perks like meeting minor celebrities such as the goalie for the Montreal Canadiens of the 1970s, but it also meant the adoption of "an entirely new mindset" about his research and the work of academics. Writing for non-specialists meant he had to articulate the field's accomplishments, resolve contradictory scholarship, and contextualize compelling research—all skills that helped his scholarly writing. *Words and Rules*, he wrote, "is written as a trade book, but I would not have done it all that differently if I had written it as an academic book."[47] He wanted, at heart, to write not only for other specialists or college students, but for everyone interested in science. And his colleagues supported him, as did MIT, which he said did not view his popular writing "as some kind of shirking of professional duties." As the chart shows, media attention to Pinker largely came as he published his popular books.[48]

As his career developed, Pinker had to take care to manage his scholarly work and his public intellectual work. "That balance is something I deal with day to day: how to allocate and limit time and attention to different modes of intellectual work. At any one time, I directly supervise one to two graduate students. I publish, I'd say, one to two substantive articles in peer-reviewed scholarly journals every year, in addition to reviews, replies, and commentaries. I keep that part of my portfolio from shrinking to zero," he said in 2014, "even if it is less than what it used to be. There were times when I had five graduate students and two grants and published four or five peer-reviewed articles a year."[49]

REPRESENTING A NEW INTELLECTUAL ERA

With his next book, Pinker reinvigorated a centuries-old debate: nature versus nurture. *The Blank Slate* (2002) was a five-hundred-page assault on the Standard Social Science Model. He assailed the intellectuals who denied the reality of human nature, characterizing them as an alliance of social scientists, progressive educationalists, Marxists, liberal writers, and postmodernists who combined to suffocate intellectual life. They created what he called an Official View that babies were blank slates, their characters and attitudes shaped by experience and education. So pervasive was this argument, he wrote, that even the moderate position that biological pressures powerfully shape human nature was taboo. The Official View denied human nature and, he wrote, "distorted the study of human beings, and thus the public and private decisions that are guided by that research."

Challenging this view are what Pinker calls "the new sciences of human nature." Cognitive science, neuroscience, genetics, and evolution together point to a human mind hardwired from birth. Humans share biological directives to mate, breed, eat, seek status, forge tools and weapons, speak languages, adopt common facial expressions, gossip, and decorate our bodies. Pinker said explanations of human behavior should be rooted in this knowledge, as should political and social policies that constrain and corral

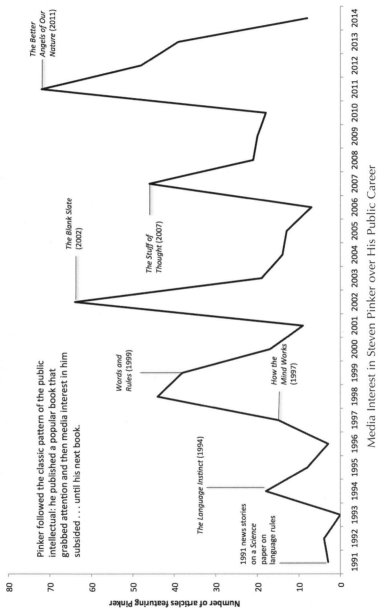

Pinker followed the classic pattern of the public intellectual: he published a popular book that grabbed attention and then media interest in him subsided . . . until his next book.

The Better Angels of Our Nature (2011)

The Blank Slate (2002)

The Stuff of Thought (2007)

Words and Rules (1999)

How the Mind Works (1997)

The Language Instinct (1994)

1991 news stories on a *Science* paper on language rules

Number of articles featuring Pinker

80 70 60 50 40 30 20 10 0

1991 1992 1993 1994 1995 1996 1997 1998 1999 2000 2001 2002 2003 2004 2005 2006 2007 2008 2009 2010 2011 2012 2013 2014

Media Interest in Steven Pinker over His Public Career

that designed behavior. But these insights—he argued repeatedly—do not threaten liberal values. Nor do they provide a biological justification for social or gender discrimination. These new sciences, instead, lead society to an alignment of science, culture, and politics in a way that merges human-ism with up-to-date scientific knowledge to form "a realistic, biologically informed humanism."[50]

The clash of ideas about the blank slate is also a clash of ideas about human nature that underlie political philosophies. The blank slate has been tied historically to progressive ideas about an equal society, hence its championing by scientists who were active in the civil rights movements of 1960s America. If we are all blank slates, there is no basis in human nature for inequalities such as sexism, racism, or economic disparities. But if there are substantial genetic differences, this is an opening to discrimination and a justification of this discrimination as a natural manifestation of innate differences in human nature. As a consequence, as the philosopher Colin McGinn noted, a scientific question becomes "charged with political meaning."[51]

Pinker avoided advocating a political position. Political controversies, he wrote, involved tradeoffs between values, and science can identify, but not resolve, these compromises. "If I am an advocate," he wrote, "it is for discoveries about human nature that have been ignored or suppressed in modern discussions of human affairs."[52] He identified himself as a political disciple of the seventeenth-century English philosopher Thomas Hobbes and, as such, believed human nature has remained unchanged. Family, re-ligion, government, and other social traditions allow people to work with a human nature that has limited virtue.[53]

But reviewers on both sides of the Atlantic said *The Blank Slate* articu-lated conservative ideas. The *Independent on Sunday* noted that Pinker's book is "one that intelligent conservatives ought to know well."[54] The *New York Times* concluded it is a calling card for compassionate conservatism.[55] Yet reviewers for liberal American political magazines praised the book. "The more we understand our nature," wrote science writer Steven Johnson in the *Nation*, for example, "the better we'll be at nurturing."[56] Philosopher and cognitive scientist Philip Gerrans noted that Pinker's conclusion is similar to Peter Singer's in *A Darwinian Left*: the propensity to violence, selfishness, promiscuity, and competition are part of our nature, as are abilities for at-tachment, nurture, and selflessness. Social institutions protect society against lawlessness, totalitarianism, and massive inequality.[57]

The blank slate, Pinker argued, caused parents avoidable grief, guilt, and anguish. All parents want their kids to turn out well, but their behaviors had little or no impact on their child's personality. Genes were much more decisive, as were kids' interactions with other children. But childcare policy is anchored, he said, on blank slate–type research that discounts genetics and concludes that parents shape how their child turns out. As a result, he wrote, "much of the advice from the parenting experts is flapdoodle."[58]

Child psychologist—and regular media commentator in Britain—Oliver James called Pinker's views "misleading," "immoral," and "dangerous," because parents who believe they have little influence are more likely to abuse or neglect their children. Characterizing the disagreement between the scientists, the *Observer* said it was more than a heated discussion about parenting, but was an "extraordinarily angry row [that] reveals the depth of the scientific battle that is emerging over the soul of mankind."[59]

When *The Blank Slate* was published, journalists returned to established ways of describing Pinker. "Pinker is not the fire-breathing kind of revolutionary," wrote the *New York Times*. "He has a thick mop of curly brown hair, edged respectably with gray, and a mild, almost diffident manner."[60] "He's a handsome man with high cheekbones and a shock of blond and grey hair worn in a style that works equally well for Led Zeppelin frontman Robert Plant," wrote the *Financial Times*. "He's dressed in worn cowboy boots and khakis. His shirt is a short-sleeved button-down, of a colour that's arguably lavender. His eyes glitter."[61]

The Blank Slate was a best seller, staying on the *New York Times* nonfiction best seller list for several weeks, sharing the space with former New York mayor Rudy Giuliani's *Leadership* and liberal filmmaker Michael Moore's *Stupid White Men.*

Critics offered reasons for *The Blank Slate*'s popularity. Eminent philosopher of science David Hull argued in *Nature* that Pinker presented "an overarching view of the world in a way that quite a few readers will find seductive."[62] Philosopher Simon Blackburn in the *New Republic* said that, despite its weak writing about culture and history, the book sold so well because it offered the "promise of a new synthesis, a science of the mind that finally tells us who we are, what is possible for us, how our politics should be organized, how people should be brought up, what to expect of ethics—in short, how to live."[63]

Louis Menand, writing in 2005 about the gulf between the sciences and the humanities, placed *The Blank Slate* in a contemporary culture imprinted with an unquestioned faith in two ideas. The first is that human behavior can be understood in biological terms. The second is that society can be best described in the language of classical liberal political and economic theory, leading ultimately to a narrative where the past is viewed as a ratification of the present.[64] Based on Menand's astute analysis, Pinker's work sits within and reflects both ideas.

The way his ideas reflected and shaped modern science meant Pinker, for science writer Robert Wright, was a landmark figure in the history of science. "Every half-century, it seems, an eminent Harvard psychologist crystallizes an intellectual era," he wrote in *Time*. At the close of the nineteenth century, it was William James. In the midtwentieth century, it was B. F. Skinner. At the opening of the twenty-first century, it was Pinker. He became the emblematic figure who came to represent an era when evolutionary psychology swept aside decades of social-science tenets. For Wright, Pinker was "on the forefront of an intellectual sea change."[65]

THE STUFF OF PINKER

Because he was celebrated by the public and respected by his peers, Pinker was an academic star, a valuable commodity for universities. They bring prestige, publicity, and funding. As part of his former tenure as Harvard University president, Lawrence Summers aimed to recruit top scientists and academic stars. Pinker was both, and in 2003 Harvard poached him.[66] Pinker said MIT suited his early-career interest in language, but Harvard was a better place to study his interest in human nature. The *New York Times* included Pinker in an article on an intensified phenomenon on U.S. campuses: the rise of the star professor. It said: "Known as the 'rock professor' for his long hair and easy style, he uses cartoons, videos, music and poetry to enliven lectures. He closed his first class by quoting Hamlet and opened another with 'If I Only Had a Brain' from *The Wizard of Oz*."[67]

Human nature was the focus of his next book, *The Stuff of Thought* (2007). It concluded two trilogies at once, knitting together the examination of language development in *The Language Instinct* and *Words and Rules* with the analysis of human nature in *How the Mind Works* and *The Blank Slate*. Language, he argued in the book, is more than merely a means of communication. Language is a window into human nature. Language—in Pinker's striking phrase—is a fistula, an open wound through which our nature is revealed to the world.

Pinker argued that humans share a language of thought that organizes the world in terms of basic physical concepts such as space, time, force, and purpose. Distant human ancestors needed to understand and use these fundamental concepts to survive. Language bears other marks of evolution. Humans use particular swear words because these taboo terms are associated with excrement and disease that could kill. Adults use polite speech—would you mind awfully passing the pepper?—not as mere cultural niceties, but as a way to negotiate potentially hostile social interactions. The skillful use of language meant survival.

Critics, in general, praised and faulted *The Stuff of Thought* for the same reasons they liked or disliked Pinker's earlier books. David Crystal, author of The *Cambridge Encyclopedia of Language*, said it was Pinker's most insightful book yet about the theoretical workings of language (and he told how he said the same thing for each of Pinker's previous books).[68] Philosopher David Papineau said he enjoyed Pinker's evidence-based discussions of language, but found his writings on evolutionary biology too speculative.[69] Oxford language professor Deborah Cameron classified Pinker's thinking as strongly black-and-white. For her, he was "the scientific equivalent of a conviction politician, and while his certainty lends his writing clarity and force, it can also make him sound glib and overly pleased with himself."[70]

For some commentators, *The Stuff of Thought* revealed Pinker's personality. William Saletan in the *New York Times* said Pinker wanted to show human nature through the window of language. "But as he does so, one more face appears in the glass: the reflection of the man looking into it." Saletan concluded: "Pinker's nature turns out to be the book's organizing principle. The linguistic arcana, the academic squabbles, the Tom Lehrer songs, the Lenny Bruce quotations—they're all part of the tale of one man's journey to understanding human nature."[71]

When *The Stuff of Thought* was published, journalists recycled the same descriptions used over a decade earlier—the *Times* said he "looks more like the frontman of a hell-raising 1980s rock band than a scientist"[72]—but also more closely interrogated his fame. The *Guardian* wrote that Pinker "seems built for the limelight to an almost parodic degree, with his Roger Daltrey hair, prominent jawline, and fondness for jeans and leather boots." Bryan Appleyard wrote in the *Sunday Times*: "Famously good-looking with his blade-like jawline, equally famously rock'n'roll with his long, curly hair and his cowboy boots, Pinker is, along with Richard Dawkins and a handful of others, a global science celebrity."[73]

The *Guardian* also noted how Pinker was used as a means of promoting other books. His publishers had arranged for him to be the headline act at a gathering of senior buyers in the book trade, and his presence was intended to let a little of his glamour rub off on the rest of the firm's titles.[74] Pinker later described how he was brought out to add glamour and gravitas to university fund-raising dinners, describing his being served "as the bait in their fund-raising dinners, in which a high-rolling alumnus or philanthropist is wined and dined until the moment is right to pop the question of a major cash donation."[75]

THE SURPRISING HISTORY OF VIOLENCE

Pinker's background in linguistics and cognitive science means that, for him, the walls between the sciences and the humanities are porous. His next book synthesized ideas and evidence from history, archaeology, psychology, and a field its advocates call "atrociology," which examines acts of wholesale historical barbarity to examine the darkest dimension of human nature and human history: violence. The eight-hundred-page *The Better Angels of Our Nature* (2011) upended the unexamined convention that our era—following the seismic violence unleashed by Stalin, Mao, the Holocaust, and two world wars—is the most violent in history.

In contrast, he argued that even though news presents a nightly orgy of destruction, the number of violent deaths—measured in fatalities per 100,000 people—in the West has for centuries been in steep decline. Violence in families and neighborhoods and between tribes and nations all follows the same downward trend. Pinker argued the diminution of violence may be "the most important thing that has ever happened in human history."[76]

Violence declined in large part because history, he argued, was driven by a civilizing process. From the end of the Middle Ages, government, trade, technology, literacy, etiquette, human rights and, crucially, reason combined to create an environment where humans' positive instincts, their better angels of morality, self-control, empathy, and reason, reined in their negative impulses, their inner demons of predatory violence, dominance, revenge, sadism, and ideology. As Pinker wrote: "The way to explain the decline of violence is to identify the changes in our cultural and material milieu that have given our peaceable motives the upper hand."

Pinker's survey of savagery over two hundred thousand years comes to a novel conclusion to support an often overlooked argument he has made consistently about human nature. It supports his view that "moral progress is compatible with a biological approach to the human mind and an acknowledgment of the dark side of human nature."[77] He ends the book with a discussion of cognitive science, arguing that people today are less likely than their forbearers to resort to violence. No bloodshed accompanies more effective behavioral strategies such as bargaining and cooperation.

The book received widespread praise and several critics said it should be essential reading in broad culture. *Nature* called it "a remarkable scholarly achievement that deserves to be studied and debated by many social scientists, concerned citizens and policy-makers."[78] The *Guardian* concluded: "everyone should read this astonishing book."[79] It sold well, too, reaching the *New York Times* extended hardback best seller list, and the same paper named it a notable book for 2011.

The book was also derided as Whig history: a historical account that views the past through the lens of the present.[80] Cultural commentator Wesley Yang in *New York* magazine wrote that the Victorian faith in science, reason, and progress was thought to have vanished. "But here comes Pinker, arriving on the other side of a century of genocide, in an age of jihad and disaster porn, armed with the work of a few dozen academic researchers." Pinker, noted Yang, took "the old and long-ago-abandoned wine of Whig history and poured it into the new bottle of a satisfyingly hard, numerate science."[81] Philosopher John Gray believed Pinker's argument was motivated principally by a worldview: the Enlightenment's commitment to reason and humanism produced advanced civilization and fostered peace.[82]

Interviews to promote the book brought readers closer to Pinker's personal life. The *Financial Times* described his open-plan loft in a converted Boston leather warehouse as it featured Pinker sitting on a "contemporary, Danish-designed sofa" in his three-room apartment with fourteen-foot-high ceilings and exposed brick walls featuring Pinker's own photographs.[83]

In one interview to promote the book, Pinker took a journalist with the *Financial Times* to shoot photos at Boston Harbor, and the resulting article tied his photographic interests to his scientific ideas. "As an evolutionary psychologist who believes that human traits are adaptive," the journalist wrote, "Pinker is convinced that we are naturally drawn to landscape photos that capture safe, inviting environments where we might be inclined to live." The paper pictured Pinker lying on the ground of

Charlestown Navy Yard with his camera, and in its headline called him "The Shutterbug Scientist."[84]

The article illustrated how Pinker's understanding of the media evolved over his public career. He chose the site for the interview with the journalist, knowing that it featured the *USS Constitution*, a vessel that fought in the 1812 war between the U.S. and Canada, former enemies that formed an enduring peace. He said: "It was a photo op for a photo op—the story was about my hobby, photography—but we both wanted it to connect, however indirectly, with my recent book on the history of violence."[85] He has learned over time that more often than not, a cooperative, rather than adversarial, relationship with a journalist works well for each party.

On this theme, Pinker noted that as a busy academic, he does not prepare extensively—if at all—for encounters with journalists. "There are only so many hours in the day and I can't devote a lot of them to image management." But Pinker noted, too, that having a public prominence means that, to some degree, there is a "self-consciousness that you have to cultivate."[86]

A BRIDGE BETWEEN SCIENCE
AND THE HUMANITIES

By the time *The Better Angels of Our Nature* came out, Pinker had helped create what he and colleagues considered to be a new science of culture— "culturnomics." In this approach, researchers used novel data-crunching techniques to search the five million books Google had digitally archived to track how often certain words appeared over time. Doing so would reveal long-term trends about topics typically considered parts of the humanities, such as collective memory and censorship.

In one of their examples, the Nazi censorship of Jewish artist Marc Chagall can be demonstrated by comparing mentions of his name in English and German books in the decades before and after World War II. Mentions of Chagall in both languages soared after 1910 and continued to rise afterward in English. Between 1936 and 1944, however, the artist's full name appears in German books only once. Between 1946 and 1954, it appeared almost 100 times. Such culturnomic data, wrote Pinker and his colleagues, constitute a new type of evidence for humanities researchers, one that supplements, not supersedes, existing forms of analysis such as the detailed, close reading of specific texts.[87] By applying the tools of culturomics to Pinker himself, it can be seen that mentions of Pinker in books started to rise in 1990. By 2000—after he had published three popular books—mentions had soared.

In 2013, Pinker asserted that science had much to offer the humanities in a 4,300-word *New Republic* piece titled "Science Is Not Your Enemy."[88] He appropriated what he considered the ill-defined term "scientism," reframing it as a way to enhance knowledge in the humanities. Scientism was used as a scare word by neglected novelists, embattled professors, and tenureless historians who resented the fact that scientific ideas and tools could advance knowledge in the humanities. He noted, too, that humanities knowledge was still scarred by what he considered postmodernism's obscure writing and hostility to truth. With *The Better Angels of Our Nature*, the *Science* article, and the *New Republic* analysis, Pinker provided the notion of a science-infused humanities with a model, a set of tools, and a philosophy.

Clear evidence of Pinker's academic stardom came in 2013 when Oxford University Press published, for a wide readership, a selection of his scholarly papers as *Language, Cognition, and Human Nature*. After the public controversies of his popular books, the collection is a reminder, too, of Pinker's excitement as a scientist. "This article," he wrote in the introduction to one piece, "grew out of a moment of astonishment that overcame me as I was working with [a colleague] on the analysis of the neural-network model of children's acquisition of the English past tense."[89] When the press offered to publish the papers, Pinker said: "I would not have agreed to mislead book buyers or saddle university libraries with another tome of recycled articles," he wrote, "if I did not think that some of my academic papers had crossover appeal."[90]

Pinker's crossover appeal, commentators noted, relied heavily on his prose style, which has remained clear and accessible even as its subjects demanded intense reader focus. He explained his approach to writing in *The Sense of Style* (2014), drawing on research from linguistics and cognitive psychology to demonstrate how language communicates ideas. For Pinker, good prose was classic prose, a style that provides a clear window into an objective reality, a style that suits scientists, as it reflects how most of them see the world.

He also debunked ideas in a text cherished by humanities professors, *The Elements of Style*. Pinker showed that the authors of the classic text, William Strunk and E. B. White, were not trained linguists and so many of their rules and examples—about the use of passive voice, for example—are incorrect and misguided. Clear writing can have far-reaching social effects, as it can bring students into science, promote the scientific worldview, and enhance understanding of science controversies.[91]

THE SUPERSTAR ACADEMIC

Richard Dawkins is the archetypal celebrity scientific intellectual, but decades ago he left the laboratory. Steven Pinker never left. He is the exemplar of the superstar scientific academic.

A prodigiously talented researcher, Pinker first studied the mind after an intellectual revolution transformed psychology and forged the new field of cognitive science. He achieved the rare and esteemed feat of making notable contributions to two subfields—visual cognition and language—of this exciting, emerging body of modern knowledge.

After he established himself as a scholar, he became the chief public explainer of new ideas in linguistics with *The Language Instinct*, a popular book that shaped public understanding of language and evolution, a book whose publication marked the beginning of Pinker's celebrification, as journalists reported on his home, dress, and—in a trope that would often recur—his long curly hair. He was the iconoclastic face of a revolutionary science.

Darwinist ideas swept through intellectual life in the 1990s—the Decade of the Brain—and Pinker became the eloquent, combative advocate for the controversial field of evolutionary biology. His fame was also based, crucially, on the way his books articulated for readers what Louis Menand identified as a dominant modern cultural trend: that humans could be explained by biology. His image crystallized this new intellectual era.

Overall, Pinker exemplifies the new breed of celebrity academic, a public star who maintains the respect of his peers, a writer of demanding books that appeal to experts and elite readers, a crossover success welcomed not only by publishers but also by universities seeking the attention guaranteed by the glamour of celebrity. Famous in both overlapping worlds, he is the ideal modern scientific superstar academic.

FIVE

The Political Celebrity of Stephen Jay Gould

Sitting beside New York Yankees legend Joe DiMaggio on the bleachers around a small San Francisco baseball field, Stephen Jay Gould opened a box that contained his most treasured possession. It was a baseball he first held as a nine-year-old boy one afternoon in 1950, as he sat with his father close to the third base line at Yankee Stadium and watched as his dad reached up and caught a foul ball struck by DiMaggio in a game against the St. Louis Browns. Gould took the ball home, washed it over and over and mailed it to Joltin' Joe, as the star was nicknamed, who signed and returned it a month later.

"I just wanted to thank you," the now-adult Gould told DiMaggio. "I was just a nothin' nine-year-old kid and you didn't have to send it back. You must have got a hundred thousand of them every day—but you did, and I just wanted you to know how much it meant to a kid growing up." "You bet, Steve," replied DiMaggio, the man who remained for Gould one of three inspirational figures in his life, alongside his father—and Charles Darwin.[1]

The encounter was arranged and staged at the Presidio of San Francisco as part of a 1984 *Nova* documentary special about Gould. But the meeting was not a superficial attempt to show the human side of the scientist. Nor was there anything artificial about Gould's effusive thanks. Baseball was a lifelong love for him, a sport he diligently followed and wrote about for decades, a sport he noted was central to the American imagination, performing the role of pastime and source of mythical heroes.

DiMaggio remains a cultural colossus. His record 56 hits as a batter in consecutive games in 1941 endures as the greatest individual achievement in baseball. His accomplishment acquired its own moniker—The Streak—as it became knitted into the fabric of American mythology.[2] DiMaggio captivated the boyhood Gould. He became, Gould wrote, "my model, and my mentor, all rolled up into one remarkable man."[3] DiMaggio, he said elsewhere, "taught me the thing that counts is excellence."[4]

DiMaggio embodied excellence for Gould in the way he mastered the varied arts of batting, fielding, catching. He also personified perfection in his unrivaled hitting streak, because "a streak must be absolutely exceptionless," Gould wrote. "You cannot make a single mistake."[5]

Before his untimely death from cancer at age sixty in 2002, Gould led a life of extraordinary intellectual achievement and public renown. He was an illustrious Harvard professor and produced a vast body of work—22 books, 101 book reviews, 479 scientific papers, and 300 essays, a total of 902 publications—that spanned evolution, paleontology, geology, natural history, history of science, philosophy of science, baseball, and the misuse of science for social discrimination.[6] He wrote an essay every month for twenty-five years for *Natural History* magazine on a manual Smith Corona typewriter, a writing streak one former baseball player likened to the Yankee Clipper's batting streak.[7] The *Observer* called him "the DiMaggio of science."[8]

He contributed revolutionary ideas to evolutionary biology, battled creationism in courts, campaigned against racism and nuclear war, inspired cancer sufferers, engaged in acerbic debates with his academic opponents, educated and delighted readers, and inspired a generation of scientists. His distinctive contribution to the public understanding of science is that he went beyond merely explaining science to his legions of devoted readers. He took them deeper into the hearts and minds of scientists, showing in his own words that while science can learn about the natural world, it "must be understood as a social phenomenon, a gutsy, human enterprise, not the work of robots programed to collect pure information."[9] Moreover, he showed how science was deployed, often unconsciously by researchers, as a seemingly objective justification for racism, oppression, and segregation. Like Pinker, he was an academic star who contributed controversial ideas to public life.

In his almost three decades in public life, Gould was an esteemed yet contentious personality. *Natural History* called him "America's Evolutionist Laureate."

The Library of Congress in 2000 named him a "Living Legend," a man who embodied the "quintessentially American ideal of individual creativity, conviction, dedication, and exuberance."[10] Yet he has also been dismissed as an "Accidental Creationist,"[11] and a researcher whom other evolutionary biologists spurned as "a somewhat woolly thinker who had managed to punch above his intellectual weight, to fame and fortune, through sheer rhetorical fire-power."[12] All the ideas and viewpoints can be examined through Gould's distinctive persona as a star scientist, a persona that developed over decades and combined "a learned Harvard professor and a baseball-loving everyman."[13]

THE FORMATION OF A SCIENTIFIC ACTIVIST

Gould was born in Queens, New York, in 1941, the grandson of Jewish immigrants from Europe. The Holocaust claimed most of both sides of his family.[14] Harsh immigration laws passed in the 1920s barred millions of Jews fleeing prewar Europe. His parents abandoned all religious belief, but remained proud of their Jewish heritage and history. His father, Leonard, was a court stenographer and a Marxist whose beliefs were forged in the upheavals of the Great Depression, the Spanish Civil War, and the growth in Europe of racism and Nazism.[15] With his background, argued historian Louis Masur, "Gould must have internalized a great deal about social class and political ideology, social discrimination and irrational persecution."[16]

His intellectual interests were cemented in early childhood. As he would go on to tell many times in his career, when he was five, his father brought him to New York's Museum of Natural History. Once he saw the giant skeleton of a *Tyrannosaurus*, he was instantly and irrecoverably smitten by paleontology. At age eleven, he read *The Meaning of Evolution* by George Gaylord Simpson "with great excitement but minimal comprehension" after his parents mistakenly received the book from a book club; they wanted to send it back but Gould said he begged them to keep it "because I saw the little stick figures of dinosaurs on the dust jacket."[17] So consumed was he by dinosaurs that his classmates, he later revealed, cruelly nicknamed him Fossil Face.

During his undergraduate years, Gould pursued his intellectual interests and continued his family's tradition of involvement in campaigns for social justice. Between 1959 and 1963 he pursued a joint major in geology and philosophy at Antioch College in Ohio, a citadel of progressive thought and social activism with an administration that wanted students to immerse themselves in social action. Gould was a member of civil rights groups, including Students for a Democratic Society (SDS), a major organization in the New Left of the sixties.[18]

He demonstrated against nuclear weapons and campaigned against segregation, participating in a sit-in to integrate an Ohio barbershop, and once running a campaign with another American student during a year's study abroad to integrate the largest dance hall in Britain, the Mecca Locarno ballroom in Bradford.[19] As historian of science Myrna Perez wrote, "Gould and his fellow students were conscious of taking an active role in the making of American identity, of changing the path of American history, and they placed themselves into a grand narrative of a fight for equality."[20]

Yet Gould followed a scientific vocation. At Columbia in 1967, he earned his doctorate in geology with a focus on paleontology. His specialist topic was the evolution of snails. For biologists, snails are a useful species to study because their life history is written on their shells. By understanding how a snail's shell evolved, scientists could therefore unlock bigger issues in evolution.

Gould focused on the *Cerion*, the most prominent land snail in the West Indian islands, and a good group of snails to study because it had an unmatched range of forms—from tall and thin, to golf balls to square—among its hundreds of species. Gould's encounters with *Cerion* molded his instincts as a natural scientist. As he later wrote: "no scientist can develop an adequate 'feel' for nature (that indefinable prerequisite of true understanding) without probing deeply into minute empirical details of some well-chosen group of organisms."[21] Writing about the snail in his one hundredth *Natural History* column, he said: "I love *Cerion* with all my heart and intellect."[22]

During his first year as a graduate student, Gould read a recently published work of scientific history that had a profound influence on his scientific thinking. Thomas Kuhn's *The Structure of Scientific Revolutions* argued that science progressed not as an ever-increasing march toward truth, but a series of revolutions where new ideas overthrow old ones. The book, said Gould, helped him reject the traditional "progressive, add-a-fact-at-a-time-don't-theorize-till-you're-old model of doing science."[23] Before it was published most scientists wanted to add building blocks to the temple of truth. Since then, "most scientists of vision," he wrote, "hope to foment revolution."[24]

After Gould earned his doctorate, Harvard hired him in 1967 as a faculty member at the American Museum of Natural History. He threw himself into the intense life of the young academic, visiting Jamaica on field trips and publishing technical reports on his findings for other experts to expand specialist knowledge and advance his career. Others noticed Gould's prodigious appetite for work. In the late 1960s, his paleontologist colleague Niles Eldredge recalled one evening when he and his wife visited the Goulds in Cambridge. "Dinner over, the evening getting late, we went to bed, but as I was dropping off, I heard the sound of Steve's . . . manual typewriter as he wrote a review," he said. "Man, that guy could put the time in."[25]

BECOMING A DISCIPLINARY SUPERSTAR

As he started work as a young academic, Gould was frustrated by paleontology's poor reputation. Its specialists were usually housed not in universities, but in museums and their wares were wheeled out periodically to entertain the public. "I don't like to say bad things about paleontologists," Nobel-winning physicist Luis Alvarez once said, "but they're really not very good scientists. They're more like stamp collectors"—a sentiment many scientists shared, Gould felt, but did not say out loud.[26] Paleontology, furthermore, had little to offer the cutting-edge field of evolutionary biology. Paleontologists, wrote biologist H. Allen Orr, "were there to dig deep holes, not to think deep thoughts."[27]

The young paleontologist with his colleague Eldredge set out to challenge that view. One of the reasons paleontologists were viewed as having little to offer evolutionary biology was their reliance for evidence on the fossil record. But this fossil record was incomplete, patchy, and had major gaps between different life forms. It did not contain enough evidence to support the conventional view that species evolved slowly and progressively across generations over millions of years.

Eldredge and Gould turned this idea on its head. The gaps in the fossil record were not gaps in the evidence for a gradualist view of evolution. The gaps *were* the evidence for a different view of evolutionary change. They argued in a 1972 paper that evolution featured long episodes of stability that were disrupted, or punctuated, by rapid short-term bursts of species change. They called the theory punctuated equilibrium.

"Punk-eek," as it became known, made Gould and Eldredge "disciplinary superstars."[28] As he learned from Kuhn, he did not wait until he was older to offer a major theory. Reflecting on Gould's early career, biologist Orr said Gould was perhaps an unprecedented type in the history of science: "the first self-consciously revolutionary scientist—the first scientist who set out to create a revolution at least in part because he felt that the field just *needed* one."[29]

Although still in his early thirties, he was hurtling toward an illustrious scientific career. In 1973, Harvard promoted him to full professor.

The same year he was offered the chance to expand his influence outside the walls of the academy. In the fall of 1973, Alan Ternes, the editor of *Natural History*, published by the American Museum of Natural History, inquired if Gould would work for the magazine. Gould liked the idea and he proposed a column based on evolutionary biology, but one that would also explore his varied interests in history, philosophy, and the social and political influences on science. For the title of the column, Gould adopted a phrase in the last paragraph of *The Origin of Species* that Darwin used to refer to evolution: This View of Life. "Size and Shape," his first column, discussed how all organisms in their design adhere to basic laws of size and shape. It was published in January 1974.[30] It marked his transition into a public intellectual.

Among the readers of that first column was Edwin Barber, then a new editor at publisher W. W. Norton, who found it while browsing through material at the New York Public Library. "Several paragraphs in," he later wrote, "I saw a big principle made clear: in prose both elegant and reassuringly down-to-earth, it explained such curiosities as why elephants must have such big bones, and Gothic cathedrals their flying buttresses." That afternoon, he wrote to Gould and asked: "'What's a smart fellow like you doing with no books in print?'"

They met soon after. Barber signed up Gould to write a book on the racist history of intelligence testing.[31] Book publishing offered Gould a way to reach further into popular culture. In the meantime, the columns honed Gould's ability to write for nonscientists. Ternes edited the pieces to ensure they appealed more to general readers. When Gould received his first piece of fan mail, he wrote to his editor: "I think you have convinced me that the intelligent layman is not a myth, and that your magazine really does have a fascinating constituency."[32] The columns had another function: in them Gould developed his public image, a persona historian Myrna Perez characterized as "a kind of quirky and energetic polymath, ... a sage yet relatable man, filled with wisdom for the ages."[33]

His columns did more than explain scientific ideas and concepts. They revealed the real inner workings of science, showcasing his philosophy of science. "Nature is objective, and nature is knowable," he would later write in his clearest articulation of his often-described viewpoint, "but we can only view her through a glass darkly—and many clouds upon our vision are of our own making: social and cultural biases, psychological preferences, and mental limitations (in universal modes of thought, not just individualized stupidity)."[34]

A recurring theme was scientific racism: the use of science to provide a biological justification for racism or inequality. Gould's interest in this topic heightened after the 1975 publication of E. O. Wilson's *Sociobiology*. The book received generally good reviews in the months after it was published, but Gould saw sociobiology as a biological validation of social inequality, a reason to keep the status quo and not fight to change society to enhance tolerance and equality.

With another Harvard biologist, Richard Lewontin, he joined the Sociobiology Study Group, a collective of scientists who were similarly concerned about Wilson's book, a collective that outlined their objections in a much-cited 1975 letter to the *New York Review of Books*.[35] He criticized Wilson in a column, but wrote to his Harvard colleague before publication to explain he had a duty to address sociobiology, a subject that had already generated attention and controversy in the popular media, because he was the "only popularist writing a regular feature on evolution."[36] These interventions into the sociobiology debate marked Gould's entry into mainstream news.[37]

In 1977 Gould spectacularly showcased his twin roles as disciplinary star and public intellectual. That year, he published two books. *Ontogeny and Phylogeny* was a technical book about the relationship between embryonic development and an organism's evolutionary history. He melded arguments from science and from history in a manner praised in the journal *Paleobiology* as "rich in insight, in glimpses of personalities, and in evaluations of motivation and philosophy in science."[38] *Ever Since Darwin* brought together a selection of his *Natural History* essays.[39] Biologist H. Allen Orr later wrote in the *New Yorker*: "Though both received good notices, it was, more than anything, their simultaneity that turned heads."[40]

Indeed the *New York Times* reviewed both books together and praised Gould as a credible guide to the inner workings of science. "Gould not only celebrates the human imagination in science," it said, "he also insists we recognize the social and cultural influences on that imagination." The review noted that in one essay Gould reflected on the problem of how nonscientists are to judge the rival claims of experts. The reviewer wrote: "There seems to be no clear answer, but it does help immeasurably to know how science works. How to penetrate science? Start with Stephen Jay Gould."[41]

In the same edition, the paper interviewed Gould in the 1859 wing of Harvard's Museum of Comparative Zoology. Novelist Raymond Sokolov wrote that the "articulate, young (36) professor of geology" was a lifelong

Yankees fan and collector of West Indian land snails. Gould told the paper that his popular writing could harm his scientific career. "Obviously," he says, "a person like me is going to have problems. Anyone who generalizes and writes for the public, *ipso facto*, is going to be an object of great suspicion. People will say, even without reading you or knowing you, 'Oh, Gould, he's just this waffler.'" Later, the interviewer wrote: "Clearly projecting from his own anxieties, he eloquently defends his friend and 'alter ego,' Carl Sagan, as a popularizer whom other scientists ought to like, even if they sneer at his best selling account of brain research, *The Dragons of Eden*, because Sagan always proselytizes for conservative, real science."[42]

The Panda's Thumb (1980), a second edited collection, contained in its title essay Gould's emblematic example of the messy process of evolution. The curious story of the panda's thumb demonstrates that the process of evolution does not result in ideal design. The proof of evolution, rather, is best seen in the odd solutions that result from a messy, imperfect natural process. A panda is essentially a bear that evolved to eat bamboo. It seemingly has a sixth digit, a thumb that helps it hold bamboo shoots. But the digit is not a thumb. It is an elongated wrist bone that was remolded, over eons of evolutionary time, as a way to help the panda manipulate and eat its primary food. The "thumb" was unique to the particular evolutionary history of the panda. It was not a feature perfectly designed to hold bamboo. It was an odd contraption that contrived to allow the panda to better hold its food. The early pandas with a larger sixth "thumb" therefore had a survival advantage.

The Mismeasure of Man (1981) attacked scientific racism. A series of historical researchers, Gould argued, produced apparently scientific results about intelligence that justified the oppression of blacks and other classes, sexes, and races. The scientists sometimes faked evidence. In the 1940s, British psychologist Sir Cyril Burt fabricated some data to support his belief that genetics determined intelligence. His work was used in the United States by psychologist Arthur Jensen to support the view that differences in intelligence between whites and blacks were innate.

Other scientists, argued Gould, produced flawed science because of their unconscious biases. In the mid-1800s, Philadelphia physician Samuel George Morton collected more than one thousand skulls. He set out to test a hypothesis: that races could be ranked by the average size of their brains. Morton measured the volume of each skull, but a series of procedural errors meant his conclusions, for Gould, "matched every good Yankee's prejudice—whites on top, Indians in the middle, and blacks on the bottom." Morton's work provided apparently scientific justification for slavery, as whites were preordained to rule over races that were mentally inferior to them.

Drawing on these cases, Gould concluded that intelligence was not one single, heritable quality that could be measured and reduced to one number. Any claim therefore that oppressed groups—races, classes, sexes—were innately inferior was, he argued, scientifically incorrect, a malignant manifestation of biological determinism: the idea that "social and economic roles accurately reflect the innate construction of people."[43]

The book won the 1981 National Book Critics Circle Award and went on to be translated into ten languages and sell more than 250,000 copies. *Newsweek* praised it as a "splendid new case study of biased science and its social abuse."[44]

When a revised edition was published in 1996, Gould explained why claims of biological determinism recur. The reasons for recurrence are sociopolitical. The resurgences of biological determinism tend to come during political campaigns to cut government spending on social programs. "What argument against social change could be more chillingly effective," he wrote, "than the claim that established orders, with some groups on top and others at the bottom, exist as an accurate reflection of the innate and unchangeable intellectual capacities of people so ranked."[45]

But in the early 1980s, Gould became enmeshed in another controversy involving science and politics: the reinvigoration of creationism.

THE ANTICREATIONIST AND ACCIDENTAL CREATIONIST

In 1980, Gould lamented to colleagues: we now have "a creationist in the White House."[46] The creationist for Gould was Ronald Reagan. In his successful 1980 election campaign, Reagan disowned evolution as part of a political strategy to unite religious conservatives and big business under a reunited Republican party. As a result, creationists gained political influence. When evolution was taught in public schools, they wanted equal time given to their views: that the Bible was literally true and the Earth cannot be more than ten thousand years old. For Gould, their project was political: "its exploiters and fundraisers are right-wing evangelicals," he wrote, "who advance the literalism of Genesis as just one item in a comprehensive political program that would also ban abortion and return old-fashioned patriarchy under the guise of saving American families." [47]

Gould directly fought this political program. He testified in the historic *McLean v. Arkansas* trial over a law that required public schools in the state to give evolution and creationism balanced treatment. Creationists hoped Arkansas would be the first of several states to adopt the law. A lawsuit brought by the American Civil Liberties Union and filed by a coalition of parents, religious organizations, and scientists—led by Methodist minister Reverend William McLean—argued the law was unconstitutional, as religious issues should be kept out of the classroom. Gould testified in a federal courtroom in Little Rock to report the case. His deposition, taken weeks before the trial, showed his spiky exchanges with lawyers for Arkansas.

Q: Have you ever taught a course on creation-science?

A: I couldn't. There is no such thing.

Q: Have you ever discussed the subject of creation-science in your classroom?

A: Only in brief allusions in my science B-16 course [an overview of the history of the Earth and life].

Q: Do you recall what your brief allusions would consist of?

A: They were negative.

Q: I would expect that, but do you recall the content?

A: Come up to Harvard next week. I am going to give two lectures on Monday and Wednesday, which is the first lecture I will give on the subject. They mainly [consist] of attempts to show that by the definitions of science, creationism [does] not qualify. [48]

On January 5, 1982, Judge William Overton ruled that creationism, or creation science, was not science. It was religion, so it had no place in a science classroom. The case was the first and only time in American history that saw full testimony about creationism given in court. "I feel honored that I had the opportunity to help present the case for evolution as natural knowledge, and for creationism as pseudoscience," Gould later wrote, "in the only legal venue ever provided to experts in the relevant professions throughout this long and important episode in 20th century American history." [49]

Other experts praised Gould's efforts. "A person as important in science as he was thought it worthwhile to get involved," said philosopher Barbara Forrest, a critic of creationism. "He lent his reputation to get the attention of the media. He did what I wish more scientists would do."[50]

At the same time, Gould's critics said he gave unwitting support to creationists. At issue was punctuated equilibrium. The theory challenged core ideas of neo-Darwinism and so some evolutionists feared Gould's promotion of punk-eek conveyed to the wider public that the idea of evolution was unsound. Scientists were worried because the general public was already divided about evolution: a 1982 Gallup poll showed 44 percent of Americans surveyed believed God created humans in their present form.[51]

Richard Dawkins summarized scientists' fears as he described the common creationist propaganda technique that turned any minor critique of Darwinian evolution into a view that evolutionists were deeply divided over their foundational theory. "Eldredge and Gould don't whisper, they shout, with eloquence and power!" he wrote. "What they shout is often pretty subtle, but the message that gets across is that something is wrong with Darwinism." Punctuated equilibrium, he concluded, "gave abundant aid and comfort to creationists and other enemies of scientific truth."[52]

A *New Yorker* essay would later summarize these criticisms when it called Gould an "accidental creationist."[53] As historian Myrna Perez Sheldon wrote, creationism and punctuated equilibrium came to be connected in the public mind. "The controversy with creationism," she wrote, "fueled Gould's increasing celebrity to audiences across America."[54]

THE CELEBRIFICATION OF A PALEONTOLOGIST

This developing fame was clear in March 1982 when *Newsweek* put him on its cover. The magazine pictured Gould holding a fossil before a picture of a prehistoric scene. A tyrannosaurus and other dinosaurs waded through a tropical swamp. A pterodactyl flew overhead. A 4,000-word profile discussed his frenetic schedule, describing how in one month he interrupted his fieldwork in the Bahamas to fly to New York to receive the National Book Critics Circle Award for *The Mismeasure of Man*, delivered three papers to a major conference, flew to Chicago to discuss his MacArthur Foundation grant, wrote a *New York Times* essay on the Arkansas trial, wrote his regular column, and gave the last few lectures of his popular Harvard class.

Gould, the magazine wrote, "helped transform the entire landscape of evolutionary theory. His chief weapon is his uncanny knack for communicating ideas simply, elegantly and persuasively." It noted that his *Natural History* writing gave him "a power over popular opinion exceeded only by those scientific immortals who have their own series on public television."[55]

The piece also revealed his personal life. It carried a picture of Gould and his son Ethan, then aged eight, who accompanied his father on a field trip to collect data on *Cerion*. It said Gould usually sleeps between about 2:00 AM and 6:30 AM. "What time is left is jealously guarded for his family. He is close to his widowed mother, who runs a shop on Cape Cod where she sells driftwood sculptures of owls and a small stock of books, the collected works of Stephen Jay Gould," the journalists wrote. "He has attempted to shield his wife, Deborah, and their two sons from the growing publicity attached to his name. A passage in his most recent book revealed that his older son, Jesse, 12, suffers from a learning disability. A friend speaks with awe of Gould spending several hours each night patiently reading and talking to his son, never despairing that he could overcome this problem, like he has so many others, by sheer will and effort."[56]

Months after the piece appeared, Gould would see his will and effort tested. In July 1982, after a routine physical examination, Gould was diagnosed with a rare cancer, abdominal mesothelioma, an asbestos-related condition that forms in the abdomen or lung. He had a predicted eight months to live. His doctor advised him not to read about it.

As soon as he could walk, Gould went to the Harvard medical library and read about it. He used his statistical training to delve deeper into his projected lifespan, an estimate based on the average length of life of other patients with the terminal condition. Gould reasoned he was not weak, infirm, or old—he was then aged forty—so he might survive longer than the average. Reflecting on his understanding of statistics, he later wrote: "I am convinced that it played a major role in saving my life. . . . I would have time to think, to plan, to fight."[57]

He fought—he underwent a novel experimental treatment. He also wrote his columns, taught his classes (wearing a cap to cover his chemotherapy-induced hair loss), and gave interviews. "I still recall the shock of seeing him on a late-night TV program," wrote John Durant, a former editor of the journal *Public Understanding of Science*, who had seen Gould in person earlier that year. "At first, I simply didn't recognize the shriveled figure on the screen—in just a few weeks, he'd apparently aged by 30 years as a result of the harsh treatment he was undergoing."[58]

Gould asked his colleague Elisabeth Vrba to visit him and discuss research ideas. "We thought, argued, and made notes almost continuously for two days, during which he hardly ate or slept. In the midst of our theoretical discussions, he frequently rushed to the bathroom to be violently ill," she recalled. "And each time he came right back to pick up once more the threads of connection between seemingly disparate biological processes and phenomena from linguistics, philosophy, and history. I found myself forgetting how gravely ill he was, and simply felt, as always, the sheer delight of exploring conceptual issues with him."[59] Gould recovered.

Gould's cancer and his recovery were featured in a profile the *New York Times* published in 1983. James Gleick, then an editor at the paper, who went on to become a renowned science writer, discussed the history of the diagnosis, accidentally revealed during a prostate exam, reporting: "Even now, as the day wears on, the pain sometimes gets to be too much, and Gould goes off to give himself an injection of a narcotic painkiller." The profile touched on other aspects of Gould's private life. He sang in the Boston Cecilia Society. He was frequently part of the substantial contingent of Yankee fans at Fenway Park, home of their great rivals, the Boston Red Sox. It noted how he often started graduate seminars reading his hate mail. But the profile also zeroed in on a paradox in Gould's public image, one he and his colleagues knew about: he was at once "the foremost exponent of Darwinism in our time," but he also put himself more and more at odds with the orthodoxies in evolutionary theory.[60]

Gould continued his political activism in the early 1980s. The Cold War raged and he was one of a group of scientists, spearheaded by Carl Sagan, who warned of an unforeseen result of an atomic war—nuclear winter. In their scenario, warheads rain down on the Northern Hemisphere. Massive smoke clouds from burning cities and forests shroud the Earth, blocking out the sun and leading to months or years of freezing darkness that decimate life on Earth. Their vision of a nuclear winter was based on a historical precedent: the catastrophic impact of an asteroid or comet that smashed into the Earth sixty-five million years ago. The impact spawned a huge dust cloud that shut out the sun and led to the extinction of the dinosaurs. One historian of science described the efforts of prominent scientists to raise awareness of nuclear winter as "a sophisticated publicity campaign."[61]

On September 12, 1984, Gould appeared alongside Sagan at hearings the U.S. House of Representatives arranged to raise the public profile of nuclear winter. Committee chairman James H. Scheuer introduced Gould as a scientist "widely recognized as one of the most seminal evolutionary biologists and as one of the best scientific communicators of our time."[62]

As he often did in his columns, Gould opened his testimony with a biblical analogy: He described how Moses terrified Pharaoh with a pall over Egypt, one that lasted three days. "The much longer darkness of nuclear winter," Gould continued, "with its associated set of consequences, including plummeting temperatures, increased radiation, fires, enhanced chemical pollution of atmosphere and surface waters, has made the prospect of nuclear war even more terrifying than terrified humanity had previously imagined."[63] Despite their efforts, the campaign had no clear effect on official nuclear policy. After the end of the Soviet Union, security and policy concerns switched to nuclear terrorism and proliferation.[64] Nuclear winter faded out.

The celebrification of Gould continued with the 1984 *Nova* special that brought him together with DiMaggio. The opening scene showed Gould playing catch with his son Ethan at a ball field in San Francisco. The first words of the narrator were: "Stephen Jay Gould, baseball fan and evolutionary biologist, is one of the liveliest voices in science today." The documentary showed Gould hunting for specimens in a forest, lecturing in apartheid South Africa on science and race, testifying on nuclear winter, and meeting Joe DiMaggio, who gave some batting tips to a Red Sox uniform–wearing Ethan Gould.

It also showed him teaching the first meeting of his general education Harvard class on the History of the Earth and of Life. Science, he told his 300 students, "creates culture by instigating change through its discoveries, but it also reflects culture because it's done by human beings who are enmeshed in the biases and thoughts of their age. They're no different from anybody else. Scientists aren't special. That's one of the main themes of this course," in a view of science he said went often unacknowledged by other scientists. "Science leads and provokes change. But science is also embedded in culture and often reflects the largely unconscious biases of those who do the work."[65]

A significant indication of Gould's popular status was the fact that he was the subject of a 1986 profile in *People*. It described the "rumpled, kinetic Gould" as "an ebullient man with a near-perpetual smile," an exception to the "stereotypical man of science," a humanist, a baritone singer in Boston's Cecilia Society, lover of Gilbert and Sullivan, and devotee of baseball, especially DiMaggio. The profile also highlighted other features of his fame. "People perceive me as a commodity," he told the magazine. "They just don't think anything of asking for five minutes of my time. It never occurs to them that if they're asking for it and another thousand people are asking, I don't have 1,000 five minutes to give."[66]

The article described how some of his scientific colleagues sniped at him. "There's a great deal of resentment of him in the scientific community," it quoted David Raup, a geology professor at the University of Chicago, as saying. "People say he's glib and superficial. It's widely assumed that Steve spends most of his time looking for publicity." Yet *People* summed up his persona, saying Gould had become "a popular symbol of erudition and scholarship."

Fame impacted his teaching work. Many Harvard students took his introductory class on life's history because he was a star. The class was regularly oversubscribed, with 800 students turning up to snag one of the 300 places on offer. Like his essay writing, the course was a product of decades of devoted attention. As a former doctoral student of his recalled, "Steve is well known to have made every lecture every year, even through two bouts of cancer, as if to emulate the hitting streak of DiMaggio he so admired."[67] Gould's teaching style mirrored his writing style. In his lectures, he told stories to illustrate bigger ideas and left students with a clear conclusion.

His fame also had consequences for his status within science. A former editor of *Paleobiology* said he accepted from Gould an ornate writing style and longer article length that he would not take from anyone else. The star paleontologist, a former doctoral student recalled, made his peers proud and embarrassed. He was one of their own and made them think deeper about their subject, but he was also "sometimes, well, just a little much." Even early in his career he produced "mixtures of admiration and dismay at our own professional meetings for his intellectual and rhetorical pyrotechnics." At one talk at a major geology conference, for example, a colleague recalled how Gould "discoursed at length on hyena penises."[68]

Wonderful Life (1989) recounted what Gould regarded as one of science's greatest ignored stories—the fossils of the Burgess Shale. Located high in the Canadian Rockies, the Shale is a ten-foot-thick layer of rock about the size of a city street. It contains the best record of the array of strange creatures that existed more than 500 million years ago. The animals lived just after an evolutionary big bang in ancient oceans: the Cambrian explosion, a period of a few million years—a flash of time in geological history—that moved into motion the evolution of modern life. The animals were preserved as exquisite three-dimensional fossils, giving scientists a rare chance to observe entire organisms and learn more about how evolution worked. The fossils, wrote Gould, "are grubby little creatures of a sea floor 530 million years old, but we greet them with awe because they are the Old Ones, and they are trying to tell us something.[69]

They told us, argued Gould, that evolution was a lottery. The heroes of *Wonderful Life* were three British paleontologists who examined the fossils in the 1970s and categorized each of them into twenty-four distinct body types. Only four of these types can be found in descendants in later eras. Most were wiped out. For Gould, this showed the conventional tale of evolution was wrong. The story was not one of an increasing diversity of complex life forms. The story, instead, was one of chance or, in one of Gould's favorite concepts, contingency. The story told of "a staggeringly improbable series of events . . . utterly unpredictable and quite unrepeatable."[70] In his often-quoted metaphor, wind the tape of life back to the time of the Burgess Shale, press play—and a different sequence of events would occur. Modern life would be profoundly . . . different.

Before *Wonderful Life*, the Burgess Shale never captured the public imagination. Gould compared the lack of attention the fossils received to the massive attention received by what he viewed as the other most important modern paleontological finding: the conclusion by Nobel-winning physicist Luis Alvarez that a meteorite caused the dinosaur extinction. Why the difference in attention? For Gould, there were two reasons. First, the scientists who studied the shale, painstakingly categorizing its odd creatures, did not conform to the stereotyped image of the scientific method: "a man in a white coat twirling dials in a laboratory—experiment, quantification, repetition, prediction." Second, the paleontologists conducted a type of work that the public viewed as inferior to physics at the apex of the sciences. "The impact theory," wrote Gould, "has everything for public acclaim—white coats, numbers, Nobel renown, and location at the top of the ladder of status."[71] *Wonderful Life* by 1992 had sold fifteen thousand hardback and thirty thousand paperback copies. It won the Rhône-Poulenc Science Book Prize. One of the paleontologists whose work Gould drew on, Simon Conway Morris, challenged Gould's interpretation in his own book, *The Crucible of Creation* (1998). Gould's reputation brought the Burgess Shale into the public eye.

Books continued to pour out of his beloved Smith Corona typewriter. In the 1990s, his best *Natural History* essays were collected in *Bully for Brontosaurus* (1991), *Eight Little Piggies* (1993), *Dinosaur in a Haystack* (1995), and *Leonardo's Mountain of Clams and the Diet of Worms* (1998). "As millions of copies sold around the world," wrote his editor at W. W.

Norton, Edwin Barber, "Steve's books found a public eager for not only more of Gould but also more of science in all its variety."[72]

The *Independent* said since he published his first collection, "the series has grown from a cult into an industry, and the collected versions of essays have sold in phenomenal numbers."[73] On the U.S. lecture circuit, he was much sought after and often drew thousands to listen to his discussion of evolution.[74] It was later stated that Gould earned $300,000 a year just from speaking engagements, and he ordinarily earned a seven-figure salary.[75] The *Observer* called Gould "a Western publishing phenomenon."[76]

Reviewers of the books went beyond describing Gould's ideas. Essayist and critic Phillip Lopate, for example, produced one of the most insightful analyses of the scientist's crafted public image. "Like all self-conscious essayists, he has erected a persona—and first-person stand-in—that selectively mirrors the author in real life," he wrote in the *New York Times*. "Mr. Gould's 'I,' as he chooses to represent himself, is first of all a mensch: the public-school Jewish kid from Queens, a 'dinosaur nut' (ridiculed as 'Fossil Face' by his schoolmates), who became a lover of tolerance and reason, a 'humanist at heart,' a 'meat-and-potatoes man.'" He is a self-promotional researcher, a polemicist for Darwinism, "a melancholy, introverted antiquarian" and "extroverted lecturer"—images that are layered above his presented image as an "earthy Everyman."[77]

In 1997, Harvard released an official photograph of Gould, which shows him in the Harvard Museum of Comparative Zoology, amid his large library, standing between an exhibit showing skulls at various stages of evolution and a sheet of music. He is in his usual state of casual dress: white undershirt, button-down blue shirt, sleeves rolled up, glasses around his neck, and pens and notebook in his shirt pocket. He presented the image of the eminent scholar as ordinary Joe.

Describing an interview with Gould over a meal, a reporter with the *Financial Times* wrote: "My luncheon companion cringes when I order 'sparkling mineral water.' He prefers tap water, he proudly tells the waiter. 'It's good, it's good for you and it's free,' he explains. Stephen Jay Gould may be the world's most famous paleontologist, but he hasn't forgotten his working-class roots."

The journalist continued: "Given his obsession with American baseball, I suggest he might be tempted by the 'gourmet frankfurter,' listed on the menu at $10. But no. 'It's probably better at Charlie's sandwich shop for $3.75,' says Gould, reading my thoughts. He settles on the better value lunch of soup and a sandwich. This sort of arrogant humility is vintage Gould." At the end of the

An official photograph of Stephen Jay Gould from Harvard University in 1997 shows its most famous scientist in his office in the Museum of Comparative Zoology, with the objects of his research: fossils and books. A decade earlier, *People* magazine called him "a popular symbol of erudition and scholarship." AP Photo/Pat Greenhouse

interview, the journalist asked Gould if a photographer could come around to take his picture. He agreed, but set a limit of 15 minutes, and he added: "I don't want anything of me standing by a dinosaur."[78]

The 1990s also saw Gould subjected to more scathing criticism from British evolutionary biologists who said he misled the public about evolution. "Because of the excellence of his essays, he has come to be seen by non-biologists as the preeminent evolutionary theorist," wrote John Maynard Smith in 1995. "In contrast, the evolutionary biologists with whom I have discussed his work tend to see him as a man whose ideas are so confused as to be hardly worth bothering with, but as one who should not be publicly criticized because he is at least on our side against the creationists." For Smith, none of this would matter, were it not that Gould gave "non-biologists a largely false picture of the state of evolutionary theory."[79]

In the same decade, Gould revealed more of his personal life. *Full House* (1996) returned to his ordeal with cancer, a topic he had avoided writing about in detail, because as "an intensely private person," he viewed intimate disclosure in print "with horror." But the story of how he lived far beyond his predicted lifespan illustrated *Full House*'s thesis. The average lifespan of sufferers was eight months. Yet some with the condition died before eight months and others, like Gould, lived for years and years after diagnosis. The average was not the reality. The ultimate reality was overall variation in lifespan.

"I am one single human being with mesothelioma, and I want a best assessment of *my own* chances—for I have personal decisions to make, and my business cannot be dictated by abstract averages," he wrote. "I need to place myself in the most probable region of the variation based upon the particulars of my own case; I must not simply assume that my personal fate will correspond to some measure of central tendency."[80]

The same decade, he divorced his first wife, married artist Rhonda Shearer, and moved to the SoHo district of Manhattan. *Architectural Digest* in 1997 wrote about their 4,200 square-foot renovated downtown loft, carrying photos of the interior, describing how Gould retreated to his library to work, and revealing how the couple chose, for the seats around their dining room table, "an arrangement that characterizes the evolution of the chair from the seventeenth through the twentieth centuries."[81] He appeared as himself that year on *The Simpsons*.

But the most intimate disclosure came in *Questioning the Millennium* (1997), his historical account of how the timing of the millennium was calculated. The book ends with a discussion of a young man who, given a birth date, can instantly name the day of the week on which the birth fell. The young man is autistic, with good language skills, but very limited in cognition. His arrangement of time allowed him to anchor and make sense of the world. "His name is Jesse," wrote Gould in the book's last lines. "He is my firstborn son, and I am very proud of him."[82]

The end of his career saw Gould unite, in a series of books, the major themes of his life. He sought in *Rocks of Ages* (1999) to reconcile science and religion. Describing himself as an agnostic, he wrote: "I believe, with all my heart, in a respectful, even loving, concordat between the magisteria of science and religion."[83] He viewed them as separate intellectual domains, nonoverlapping magisteria (NOMA). Science in his view documents and explains the facts of the natural world, whereas religion examines questions science can never resolve around meaning and moral value.

The book, however, received some of the worst reviews of Gould's career. "It does not take an unbiased reader long to conclude that NOMA is a nonstarter, destined to plunge to the ocean floor straight from the launching ramp, with chips of champagne-bottle still on its nose. There are several reasons for this, but the most obvious is Gould's glaringly inadequate account of religion," wrote literature scholar and Gould fan, John Carey. "Gould seems perfectly serious about Noma, otherwise you might suspect that a peace-treaty that entailed such a drastic reduction of religion's armory was a sort of joke."[84]

In 2002, Gould published *The Structure of Evolutionary Theory*, the summary of his life's work, the scientific book he labored over for more than two decades. Across its 1,433 pages, he examined the history of evolutionary thought before he detailed his own interpretation of the current state of evolutionary knowledge. The book's blurb called him the "world's most revered and eloquent interpreter of evolutionary ideas." It characterized the book as a "peerless work, the likes of which the scientific world has not seen—and may not see again—for well over a century."

The book's reviews, even by critics hostile to Gould's perspective, praised it as a stunning intellectual achievement. Zoologist and science writer Mark Ridley argued that Gould tended to celebrate his supporters while "ignoring, distorting or psychoanalytically shrinking his critics." Despite this, wrote Ridley, the book was "still a magnificent summary of a quarter-century of influential thinking and a major publishing event in evolutionary biology."[85]

But philosopher of science Michael Ruse said the book was the product of ego and insecurity. The size of the book "is a testament to unrestrained ego, and page after page reinforces the theme. The science is lost in acres of verbiage about the merits and achievements of the author," he wrote. "I can only hope that—with this brontosaurus of a book—Gould has expiated the insecurities he obviously feels about his standing as a real professional biologist."[86]

ESTABLISHING GOULD'S LEGACY

"We were stunned today to learn that one of the country's most widely read and most appreciated scientists has died," veteran ABC News anchor Peter Jennings told the American nation on May 20, 2002. "Stephen Jay Gould had a uniquely diverse and interesting career." The cause of death was cancer. He died at his New York home, surrounded by his wife, his mother, and his books. *Nature* called him "the world's most renowned paleontologist."[87] *New Scientist* said Gould was "one of the most influential evolutionary thinkers of the 20th century."[88] The *Daily Telegraph* said: "For Americans, Gould's heavy-lidded eyes and bushy moustache represented the public face of science."[89] The *New York Times* said he "came to stand for science itself in the minds of many lay readers."[90]

With his death and the publication of his master work, media attention to Gould soared, as the chart shows, to its highest-ever level.[91] Writers and commentators differed on his lasting legacy. The *Independent* said Gould's contribution was that he "single-handedly returned a dry and marginalised science of paleontology into a major player in the game of evolution."[92] The *Observer* said he resurrected the scientific essay and proved the abstraction called the intelligent layperson does exist.[93] Michael Novacek, provost of science at New York's American Museum of Natural History, said: "Probably more than anyone else, he provided a contextual sense of science that was incredibly effective."[94] *BioSciences* said his scientific legacy was uncertain, but his contribution to public culture, the way he contributed to public awareness of evolution and brought attention to the Burgess Shale, was beyond doubt.[95]

The Gould industry continued after his death. *Triumph and Tragedy in Mudville* (2003), a collection of his baseball writings, was praised in the *New York Times* as describing the same principles of change and continuity that feature in his scientific work. (The review's headline read: "Professor in the Bleachers with a Lifelong Scorecard.")[96]

The Hedgehog, the Fox, and the Magister's Pox (2003) sought to reconcile the sciences with the humanities. Both ways of knowing the world were different, but equal, paths to a common goal: human wisdom. The book was poorly reviewed. The *Financial Times*, for example, said: "the book has the air of an unedited ramble that fails to do justice to the author."[97] Yet it offered a concrete example of someone considered by Gould to embody the fruitful interaction between both disciplines: the novelist Vladimir Nabokov.

The Russian writer was an expert on butterflies who worked in the 1940s as a scientist at Harvard. His numerous, detailed descriptions of moths and butterflies in his prose showcased, for Gould, the veneration of reality common to scientists and artists. "Nabokov, as one of literature's consummate craftsmen," Gould wrote, "upheld the sacredness of accurate factuality—an obvious requirement in science, but also a boon to certain genres of literature."[98]

Other publications tried to frame Gould's legacy. *Paleobiology* published a special edition in his honor. A collective of former students, colleagues, and collaborators took the role of custodians of his reputation, offering their interpretations—before biographers picked over his life—of Gould as a researcher, teacher, and friend in *Stephen Jay Gould: Reflections*

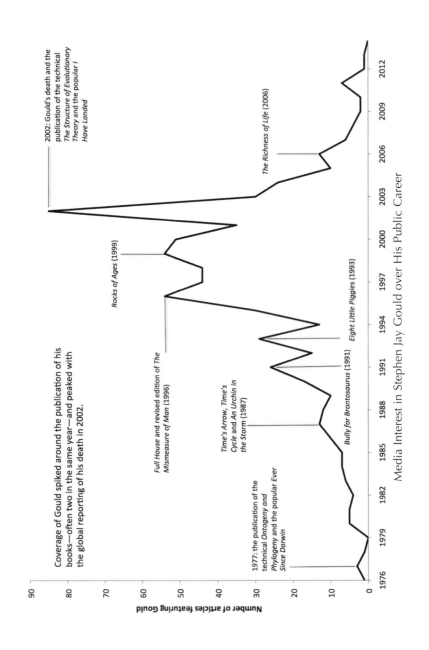

Coverage of Gould spiked around the publication of his books—often two in the same year—and peaked with the global reporting of his death in 2002.

2002: Gould's death and the publication of the technical *The Structure of Evolutionary Theory* and the popular I *Have Landed*

Rocks of Ages (1999)

The Richness of Life (2006)

Full House and revised edition of *The Mismeasure of Man* (1996)

Time's Arrow, Time's Cycle and An Urchin in the Storm (1987)

1977: the publication of the technical *Ontogeny and Phylogeny* and the popular *Ever Since Darwin*

Eight Little Piggies (1993)

Bully for Brontosaurus (1991)

Number of articles featuring Gould

90
80
70
60
50
40
30
20
10
0

1976   1979   1982   1985   1988   1991   1994   1997   2000   2003   2006   2009   2012

Media Interest in Stephen Jay Gould over His Public Career

on His View of Life (2009). The book, nevertheless, was not altogether adulatory. "He was a difficult role model," noted one contributor. "He decided quickly whom he did and didn't favor, and you usually didn't get a second chance to make a first impression."[99] *The Science and Humanism of Stephen Jay Gould* (2011) positioned Gould within a tradition of left-wing scientists who wanted science to improve society: they wanted to "build an informed citizenry that could actively engage in intellectual matters, rather than let science remain exclusively in the hands of a social elite."[100] An intellectual biography, *Stephen Jay Gould and the Politics of Evolution* (2009), attributed his popularity to how he showed a true picture of how science works, as opposed to the sterile stories readers had been fed in textbooks and dull public pronouncements.

In a contingent event that would probably have made him smile, Gould got to speak to his readers about future work—after he died. An interview he gave to the *New York Times* came to be published two weeks after he passed away. He said he hoped *The Structure of Evolutionary Theory* would be read in a hundred years. He told how Gilbert and Sullivan embodied a kind of excellence in popular music. He described the two books he planned to write next. One would be about patterns: it would be called *Life's Direction*. The other would detail the early history of paleontology. "I've been building up an antiquarian book collection for decades. I own most of the great works of 16th-to-18th-century paleontology. I can read them," he said. "So that will be the big retirement project, I think. But I'll need 20 years to do those."[101]

POLITICAL STAR

Gould once wrote that we must judge writers by their arguments, not their autobiography. But his public life shows how this ideal has vanished. In the same way as politicians' private lives now underwrite their public values, he shows how public scientists' personal histories underwrite their public images.

A descendant of Jewish immigrants, Gould's upbringing and education forged in him a passionate belief in equality and social justice, a belief that animated his scientific work and public life. He campaigned in person against segregation in an Ohio barbershop and a British dance hall. He testified in court against teaching creationism in schools, which he saw as the first salvo in a wider political campaign to instill regressive values at the heart of American life. He lectured against racism in apartheid South Africa.

He shows how political beliefs influence scientific work that is supposedly neutral. He fought in his writing against scientific claims of a hardwired human nature that he believed were periodically wheeled out in times of social crisis to justify inequalities between sexes, classes, or groups. He demonstrated that many experiments that found differences in intelligence between races were in fact based on sometimes unconscious biases and drives that gave racism a fake veneer of scientific fact.

More broadly, he demonstrated, through his own work and the many researchers he wrote about, that science had the cognitive power to understand the world, even as it was enmeshed with factors that were viewed as outside the scientific process—like scientists' politics, their social conditions, and their surrounding intellectual environment. Gould's clear expression of these complex ideas in broad culture was, for two of his closest colleagues, a vital part of his legacy. The left-wing scientists Richard C. Lewontin and Richard Levins said Gould made "a broad public think again about the validity of a Marxist analysis."[102]

Lewontin and Levins argued these views, too, throughout their prolific careers. Yet neither became a celebrity like Gould. A reason was that Gould did not come across as an aggressive ideological fighter. He was instead portrayed as an emblem of general erudition, a lovable polymath who worked in a room full of fossils, a researcher who loved snails and dinosaurs and DiMaggio. His views appealed because he appealed. Autobiography and argument fused. Gould's fame is political fame.

SIX

Fame and the Female Scientist: Susan Greenfield

On a Friday evening in 2004, a party took place at the Royal Institution (RI) of Great Britain, science's "equivalent of the theatre,"[1] situated on Albemarle Street in London's upmarket Mayfair. The RI has devoted itself since its 1799 foundation to scientific research and public communication. Fourteen Nobel laureates lived there when they received the prize. Ten chemical elements were discovered in its laboratories. Chemists Sir Humphry Davy and Michael Faraday remain former directors.[2] And it was in Faraday's original suite of apartments that the party was held ahead of a lecture by Nobel prize-winning Caltech chemist Professor Ahmed Zewail.

He had been invited to speak by the institution's then director, Professor Susan Greenfield, the first female to hold the prestigious role. Covering the event, the *Sunday Times* described how Greenfield, dressed in "floating black chiffon with draped neckline by Whistles, and the sexiest teetering ankle straps you have seen outside clubland," greeted her guests "with kisses and champagne." She had an apartment at the institution, "a stone's throw from Gucci and Prada in Bond Street, where [she] is known to shop for her designer wardrobe." As Greenfield prepared for the party, her "jumble of beautifying lotions and lipsticks [were] spilling over Faraday's elegant dressing table."[3]

Greenfield envisioned the RI as somewhere people would go "to discuss quantum physics over a latte,"[4] as a "salon for science,"[5] or "a Groucho Club for scientists,"[6] in reference to the London club frequented by artists, writers, and musicians, a destination with a reputation in the 1990s for bad behavior.

Since her appointment as director in 1998, Greenfield oversaw a £22 million redesign of the building by architect Sir Terry Farrell. Historic rooms were restored. The Faraday Lecture Theatre was refitted. Exhibition spaces were created. A bar, café, and restaurant were installed. Written on the cubicle doors of the ladies' restrooms were the chemical formulae for items such as lipstick.[7] Couples could get married within its walls.

The institution's glamorous rebranding reflected the tastes and styles of Greenfield. *Nature* called her a "celebrity neuroscientist."[8] *Science* labeled her a "science rock star" who came "alive in the spotlight."[9] Newspapers variously described her as a "self-styled celebrity scientist . . . England's most famous neuroscientist . . . arguably Britain's most well known, not to mention most glamorous, boffin,"[10] "science's most famous woman,"[11] "Britain's most famous living female scientist,"[12] "Britain's leading female scientist,"[13] "Britain's best-known female scientist,"[14] a "star media boffin,"[15] "a media don,"[16] a "mini-skirted media celebrity,"[17] and a "dolly-bird boffin of tabloid fame."[18]

These florid descriptions belie Greenfield's scientific achievements. As well as the former head of the RI, she is Senior Research Fellow at Oxford University researching conditions such as Alzheimer's and Parkinson's, the author of several popular books on the brain, the holder of a prestigious Michael Faraday medal awarded by the Royal Society for her science communication work, and a member of the UK House of Lords, where her official title is that of Baroness of Ot Moor in the County of Oxfordshire.

Beyond her accomplishments, Greenfield strikingly illustrates the controversial ways female scientists have been historically presented in popular culture. As opposed to Hawking, Dawkins, Pinker, and Gould, she represents vividly how modern research has been increasingly tied to the marketing, promotion, and selling of science. Greenfield shows that media visibility and popular reputation are valuable qualities for today's scientist. But she also shows—strikingly—how those same qualities can also lead to caustic censure from critics and ugly backbackbiting from peers.

THE ENTREPRENEURIAL SCIENTIST

Greenfield first established her scientific reputation in the late 1970s and early 1980s. She charted, in a series of papers, the novel biological functions of the enzyme acetylcholinesterase (AChE), which she argued might link several neurodegenerative diseases, including Alzheimer's and Parkinson's. The enzyme had been associated until then with development in the brain, but not with disease. Her 1984 paper on the novel functions of the enzyme remains her most-cited scientific paper, with more than 220 citations in other scholarly papers.[19]

Greenfield began her studies in philosophy, but became frustrated by what she saw as the discipline's endless discussion of language. She switched to psychology, but grew tired of the endless experiments on rats (her undergraduate thesis was on physiology, anatomy, and behavior).[20] Drawn by the promise of pursuing fundamental questions, she chose neuroscience for her doctoral work. "Because I'd come from an unusual background anyway," she would say, "I hadn't been brought up in one thing or another, I could range freely among these different disciplines, smashing through these different barriers, not having the complete definitive expertise in any one."[21]

As Greenfield's career progressed, science in Britian became more commercialized, as part of the wider privatization and free-market agenda associated with Conservative prime minister Margaret Thatcher. In 1988, Thatcher—who studied chemistry as an undergraduate—told the Royal Society that academic and industrial scientists should combine their expertise. "Industry is becoming more scientific-minded: scientists more industry-minded," she said. "It is only when industry and academia recognise and mobilise each other's strengths that the full intellectual energy of Britain will be released."[22]

Greenfield embraced this commercial ethos. At the end of the 1980s, Oxford's Pharmacology Department needed funding so she and her biochemist colleague David Smith sought private investment. They organized a neuroscience symposium for pharmaceutical company Squibb, whose representatives, Smith recalled, were particularly impressed with Greenfield. In 1987, Squibb gave the department twenty million pounds—then "the single largest university grant ever offered in England"[23]—in exchange for intellectual property rights to the work of its neuroscientists.

Moreover, at the end of the twentieth century, cognitive neuroscience became a fashionable field. In 1990, President George H. W. Bush proclaimed the 1990s the "Decade of the Brain." He explained: "The need for continued study of the brain is compelling: millions of Americans are affected each year by disorders of the brain ranging from neurogenetic diseases to degenerative disorders such as Alzheimer's, as well as stroke, schizophrenia, autism, and impairments of speech, language, and hearing."

The president concluded. "[T]hese individuals and their families are justifiably hopeful, for a new era of discovery is dawning in brain research."[24]

Greenfield's first appearances in media came in reviews of a book she coedited in 1987, *Mindwaves: Thoughts on Intelligence, Identity, and Consciousness*. The collection of 32 scholarly essays on consciousness by neurobiologists, philosophers, linguists, physicians, and computer scientists signaled her interest in an interdisciplinary approach to fundamental questions of identity and the self. The *Times* praised the book as "a serious and well-argued collection" that like "the field it deals with . . . is a sort of intellectual jig-saw puzzle of converging or conflicting arguments."[25]

Greenfield also reviewed books for *Nature* on topics related to the mind.[26] The editor of one of her *Nature* pieces also worked for the BBC and needed someone to speak about women in science for a late-night television show. He contacted Greenfield. *Current Biology* would later report: "the editors were well pleased with what they heard and saw."[27] Other scientific elites noticed her. As science writer Ted Anton would later observe of her late-night BBC appearance, "[a]ttractive, quick-witted, she caught the attention of the Royal Institution."[28]

A PIONEERING FEMALE TV SCIENTIST

The RI in 1994 chose Greenfield as the first woman to give its annual series of Christmas lectures for children: the prestigious lectures have long been a fixture of the British scientific calendar, with previous high-profile speakers including Richard Dawkins, David Attenborough, and Carl Sagan. Her talks were straightforward explanations of the workings of the physical brain. They marked the most significant milestone so far in Greenfield's emergence as a public scientist.

Journalists discussed her pioneering role as RI lecturer. They also focused on how she looked. The *Independent* noted that, in addition to her expertise, her "infectious enthusiasm, the striking outfits she favours, her brown eyes, blonde hair and wide smile will not have gone unnoticed."[29] The *Times* wrote that Greenfield wore "a blush pink silk blouse and leggings to bring a ridiculously long tradition of bow-ties and tweed-jackets to its overdue end." The paper's writer reviewed the lectures alongside another show that featured *Playboy* founder, Hugh Hefner. Tying together both subjects, the reviewer said: "Dr Susan Greenfield may not have been wearing a Bunny costume, but she might as well have [been] for all the fuss that has surrounded the first woman in almost 170 years" to give the lectures.[30]

The same reviewer called her "Renaissance Woman personified."[31] Profiling her, the *Times* called her "one of the brightest of her generation."[32] Greenfield herself wrote in the *Independent*: "Even though I don't want to be judged as the first woman, I know that I will be, and so I want to make a really good show of it."[33]

Journalists described Greenfield as exceptional, a common pattern of media representation where female scientists were portrayed as token figures. Coverage depicted women as exceptional figures, who overcame hardship to earn achievements that were perceived to be unique and almost peerless. These token figures were often "cultural heroes."[34] Women researchers, for example, were largely invisible on American television, and the dramas and documentaries that did feature female scientists portrayed them as stereotypical superwomen consumed by their careers, or as "romantic, adventurous celebrities like Margaret Mead and Jane Goodall."[35]

The lectures, nevertheless, led to greater media exposure. "By the end of the year I was getting all sorts of invitations, writing for the papers and doing local radio interviews. I've been on *Tomorrow's World*, *Question Time*, *Any Questions*, *Desert Island Discs* and *Start the Week*."[36] With appearances on these prominent shows in Britain, Greenfield entered the cultural mainstream.

POPULARIZING THE BRAIN

Her newfound cultural prominence led to *Journey to the Centers of the Mind* (1995), a popular book about a contentious topic: the nature of consciousness. For her, consciousness was created by groups of tens of thousands of neurons firing together in what she called large assemblies. The assemblages of neurons firing together could be studied as an index of consciousness, but these neurons were not consciousness itself. Instead, consciousness was analogous to ripples in the brain spreading out from these assemblies of neurons. The physical basis of consciousness was an acknowledged area of research interest for Greenfield, and in 1996 she was appointed professor of synaptic pharmacology at Oxford.

She continued to theorize about neuroscience and consciousness in popular books. *The Human Brain: A Guided Tour*, first published in 1997 and aimed at "not just non-biologists but non-students,"[37] speculated about how the mind could arise from the physical brain. Greenfield wrote: "These ideas are not intended to be taken as hard facts but rather to excite readers into an active line of questioning and thought of their own."[38] The public understanding of brain science was of social importance too, she noted, as the pressures of contemporary life have led to an increase in mental illnesses, which included depression and anxiety and mood-altering drug dependence, as well as brain disorders among the elderly.[39]

Her intellectual range widened as she undertook the role of the public intellectual. Between 1996 and 2001, Greenfield was a columnist and contributor to the *Independent* and the *Independent on Sunday*. Her pieces examined issues that affected the professional lives of contemporary scientists, but she also addressed the public roles of scientists. Controversial science must be debated, she argued, criticizing the distinction between researchers who were "ivory-tower purists" who did not communicate with citizens and "the handful of very talented, full-time professional publicists" who did.[40]

The characterization of the science communicator as science publicist was used—sometimes pejoratively—to describe Greenfield herself: she would later be called variously a "promoter extraordinaire of science," "the motormouth publicist of science," an "academic glamourpuss"[41] and "an evangelist of science for all."[42]

But none of her colleagues commented on her newspaper columns. "They never mention it, never mention any of my activities. From which I interpret a lot of bitching goes on behind my back," she wrote in 1988. "It doesn't bother me—though I'd much rather they came up to me and said we think you're an old tart for doing it—but I do find it sad that people don't support public understanding."[43]

Greenfield wrote that she conformed to the idea of the "entrepreneurial scientist," who was comfortable accepting private money in a culture of dwindling public funding for science. There was an ideological as well as pragmatic philosophy behind her views, as she argued that public funders did not invest in original scientific ideas. "In my experience," she wrote, "mould-breaking, paradigm-shifting ideas stand no chance with the cautious public sector grandees." Funding institutions had developed a mind-set that favored the funding of what she called "safe-science" with an outcome that was assured. But, to Greenfield, the idea of a guaranteed outcome in science was paradoxical. The private sector, by contrast, had to invest in originality.[44]

As Greenfield advocated an entrepreneurial science in her public writings, she drove the commercialization of her own research. Funded by bankers, venture capitalists, and Oxford University, she co-founded in 1997 the company Synaptica to commercialize research on acetylcholinesterase. The business world viewed Synaptica as a new model for commercialized science. A 2001 report by global business consultants Pricewaterhouse-Coopers on the future of the pharmaceutical industry said that Greenfield had taken out a patent on a molecule that could help create treatments for Alzheimer's and Parkinson's diseases. Synaptica was set up to research the molecule "before selling the results to a pharmaceutical operation." Discussing Greenfield, the report said, "The industry needs just such people. We predict that a growing number of companies will therefore adapt the model for themselves" and that scientists will "ultimately have a financial as well as a personal stake in the molecules they are studying."[45]

Her strongly pro-commercial ethos merged with her public profile when she advertised and promoted MindFit, a computer game marketed at keeping the brain active. A report of the product's launch at the House of Lords described Greenfield being there, surrounded by male scientists. It said: "It's clear that they need her, these men, to sell their research and their institutions. But does she need them?"[46]

THE FIRST FEMALE DIRECTOR OF
THE ROYAL INSTITUTION

By this time, Greenfield had reached a peak of the British scientific establishment. She was appointed, in 1998, the RI's first female director. At the time, the historic organization, she would later write, had "fallen into a genteel slumber."[47] Journalists observed that her selection was influenced by considerations of modernization and marketing. *Nature*, for example, headlined its report: "Highbrow 'Club' Seeks the Common Touch."[48]

In her inaugural speech, Greenfield outlined her vision for the institution's social role. The RI should "speak with authority on scientific matters . . . to help promote science and to encourage a proper, responsible and informed approach to scientific issues," she said. "We need to seek new audiences, reach new readership. The survival of science, and our success as a nation depends on it."

Greenfield conceptualized science communication as similar to marketing. "We have a product. Our product, which we should go out and sell, is science as a concept, scientific knowledge as an integral part of our modern way of life," she said. "We are going to be looking at ways of promoting this product to a market which I know from my own experience of publishing, radio and television is crying out for it."[49]

This outlook echoed the political ethos of the time. The then British prime minister Tony Blair told the Royal Society in 2002 that a "confident relationship" between science and the public could lead to Britain becoming "as much of a powerhouse of innovation—and its spin-offs—in the 21st century as [it was] in the 19th and early 20th century."[50]

Journalist Sean O'Hagan connected Greenfield's promotional outlook with those of Blairism, a political philosophy that viewed news management and image-making as core parts of electioneering and governance. He wrote in the *Observer*: "Greenfield is the epitome of a New Labour success story . . . in her rebranding of herself, and in her spinning of events wherein the private and the public merge uncomfortably."[51]

But at the end of the twentieth century, British science and policy elites were concerned with what they saw as a breakdown in public trust in science.[52] The scientific and political controversies over so-called mad cow disease, genetically modified organisms, and claimed links between the measles, mumps, and rubella (MMR) vaccine and autism conspired with other incidents to undermine public support of science. In response, policymakers called for a new style of communication to restore public trust in science and involve citizens in science policy.

As part of a new strategy to restore trust, the influential House of Lords recommended in a 2000 report that scientists improve their media relations' skills. In response, Greenfield provided resources for the formation of the Science Media Center in 2002, an independent organization, based at the RI, which aimed to improve the quality of science coverage in the media. The center, she said, would be "unashamedly pro-science" and would "help renew public trust in science." After the political controversy over genetically modified foods in the UK, the job of the center, according to its director, Fiona Fox, was to "ensure that never again will the UK have a national media debate about an area of science without the best scientists taking up their rightful part on that debate."[53]

A POSTER GIRL FOR SCIENCE

Her pathbreaking appearance as the first female to give the Royal Institution lectures and her position as the first female director of the RI meant she became "a poster girl for science in the UK"[54] and "the public face of gender issues in science."[55] She embraced the emblematic role, identifying her influences as the iconic historical female scientists Marie Curie, Rosalind Franklin, and Dorothy Hodgkin. In her columns, she contextualized her own career within the structural constraints that affected women in science, such as the difficulty of attracting young women to science, the presence of glass ceilings in science, and the problem of women scientists having children in their mid-thirties, a time when scientists usually gained an established academic post.

Her emblematic stature translated to political influence. In 2002, Greenfield investigated the issue of women in science for an official UK government report. Women scientists, the report concluded, felt disadvantaged by "informal practices including rumor, gossip, sarcasm, humor, throwaway remarks and alliance building" by male colleagues.[56]

Women scientists, moreover, have been portrayed in popular culture as fundamentally different from their male counterparts. For literary scholar Elizabeth Leane, a "scientist who is a woman is then always a *woman* scientist, not simply a scientist; and as a woman scientist she cannot be disembodied."[57] The historian and philosopher of science Evelyn Fox Keller argued that in intellectual life, women have been concerned with the personal, the particular, and the emotional, while science has been presented as a distinctly masculine preserve, associated with rationality and the impersonal. Objectivity was *hard* and associated with masculinity, she noted, while subjectivity was *soft* and associated with the feminine.[58]

PUTTING THE "SEX INTO SCIENCE"

Yet Greenfield has been portrayed in stereotypical ways. While male scientists were discussed in terms of their public role, female scientists were presented in terms of their domestic, personal, and professional lives.[59] One in five newspaper profiles of male scientists referred to their appearance, clothing, physique, or hairstyle—but one in two profiles of women referred to these elements. Some portrayals also emphasized the female's sexual attractiveness to men in an effort to "sex up" science. Women were asked by photographers to adopt particular poses (sitting with legs crossed on a lab bench, for example) and were asked by television producers to dress provocatively.[60] Today, women scientists are no longer evaluated publicly on their baking or sewing skills, as they were between the 1920s and the 1980s, but "they may now be judged on beauty, fashion and sexiness." Female scientists have often been cast into these new roles, fitting with broader cultural shifts that have increased the emphasis on glamour.[61]

Greenfield was presented as glamorous. A key text was a four-page article about her in 1999 in *Hello!*, the magazine known for its uncritical and promotional coverage of social elites and celebrities, famous figures who often offer the magazine an insight into their private worlds. The article about Greenfield

Susan Greenfield poses beside a statue of iconic scientist Michael Faraday at Britain's Royal Institution, which she wanted to wake from what she called its "genteel slumber" after she was made its first female director in 1998. A slightly different version of this photograph accompanied a 1999 interview with her in *Hello!* magazine. David Montgomery/Getty Images

was accompanied by a series of photographs of her taken in the Royal Institution sitting on a desk in a lecture theater, and sitting on a couch with her then husband, chemist and popular science writer Peter Atkins. She was also pictured posing beside a bust of Faraday, an image almost identical to the one reproduced here, one that illustrates visually her attempt to modernize the institution.

The article was used as a means of promoting a glamorous and outward-looking RI. But the piece also featured personal revelation. Describing her relationship with her husband, she said: "We're close without being sentimental, best friends and soulmates."[62] The writer noted that Greenfield alternated between "sharp-cut designer suits for corporate meetings and funky silver anoraks and leggings in the lab" and she took time off her packed schedule for "dancing, eating out, and early-morning haircuts."[63]

The *Hello!* article became a point of reference for subsequent writing on Greenfield. Neglected, however, is the fact that the article promoted the public value of the RI. "I wanted to really put it on the map and promote it. It had been sinking into genteel slumber," she said in an interview. "I was trying to bring the RI back to its original roots around the sheer joy of science and not being some kind of snooty club."[64]

Journalists in various publications almost always mentioned her looks. The *Guardian* noted that she looked good in academic robes, but would "look drop-dead gorgeous in a velour shell-suit."[65] The *Guardian* also described her as "a large mouth on thin legs, somewhere between a tabloid columnist and an American intellectual. . . . She is wearing a sexy, fitted, red jacket and high, stacked heels. She is 50 this year, and looks 40."[66] The *Independent on Sunday* described Greenfield's "black knee-length boots with stack heels, a black dress and a cerise crop-top cardigan."[67] The *Independent* described her in a headline with the phrase "Lab Fab."[68]

The *Observer*'s science and technology editor wrote that, at lab meetings, "she makes a striking figure in her mini-skirt and pink Guerlain lipstick."[69] The *Daily Mail* said she was putting "sex into science," noting that "she looks immaculate in pale pink gilet, fine lilac twinset from Whistles, black pelmet skirt by Giorgio Armani, black tights and clumpy block wedges from an exclusive shop in London's Sloane Street."[70] The *Times* described her as wearing "shoes of teetering altitude, a miniskirt of dizzying brevity and flicks her blonde tresses in a manner that must make kneecaps quiver among livelier male peers in the House of Lords."[71] *Science* called her a "commandingly tall woman with leonine features."[72]

Asked about these portrayals, she said: "How many men would that happen to? None."[73]

Yet Greenfield has been active in this feminine and sexualized self-framing. She described her fashion style as "funky" and "slightly grungy" and she tied her dress sense to her work: "It's the same with science. I've always enjoyed shifting paradigms."[74] A reporter alluded to a "makeover" in 2002 when Greenfield was "still in pearls and ash-blonde layered haircuts," and the reporter noted that "unaccessorised, a sparkling intellect does not get you on to the pages of *Vogue*."[75] Another journalist noted how Greenfield "kicked up a shapely ankle to show me her new Miu-Miu platform shoes."[76]

When she was photographed by the *Sunday Times Magazine* in the Royal Institution in 2004, she wore a red evening gown and the magazine had as its front-page headline, "The Prof Wears Prada."[77] In 2000, she was reported as saying that she and Atkins shared "a respect for writing, reading and talking, and, of course, sex."[78] Referring to her *Hello!* appearance, she said she was disappointed with the photographs, as she did not look like model Naomi Campbell, but instead "like a dumpy middle-aged female scientist."[79] Elsewhere, she said: "I look in the mirror all the time. I'm vain in the sense that if I could look better in the morning and I don't, then I get depressed. . . . Of my features, I like my lips best. They're quite full, and lips can't age too much. . . . In terms of dislikes, it has to be my nose. It's too beaky."[80]

Asked what part of her body she would change, she said, "My bum. I want one like Kylie's" (pop singer Kylie Minogue).[81] She has discussed the most she has spent on a pair of shoes (300 pounds), where she had her first drink as an undergraduate (the Trout pub),[82] and has described one of her favorite restaurants in Oxford.[83] Her nickname at university was Springy. She said she cannot swim, cycle, or cook.[84] Yet in 2007, she told a journalist that she did not discuss her private life, saying: "I don't answer questions like that."[85]

The focus on her appearance and clothing has been so relentless that journalists and other writers mentioned earlier coverage when they wrote about Greenfield. Historian of science Patricia Fara noted that articles about her "invariably mention her fondness for miniskirts."[86] Sean O'Hagan in the *Observer* wondered where was her "fabled black Armani miniskirt" that she supposedly wore in the *Hello!* shoot. He wanted to see this "emblematic item of clothing, this sartorial signature" referred to more frequently in her press reports than her research on the brain or Alzheimer's.[87]

Her private life was opened to further public scrutiny after her marriage to Atkins ended. They were described as the "science world's golden couple"[88] and "science's most glamorous couple."[89] Their divorce "scandalised not only him but the entire scientific and political establishments."[90] Her marriage breakup was reported in 2003 in the gossip and diary section of the *Daily Mail*,[91] the publication in which Atkins later discussed in detail the marriage's end in an article headlined "Love, Sex and Our Marriage Split—by Britain's Brainiest Couple."[92] Atkins said that "as she became more famous and started to move in different circles in London, at the Royal Institution and in the House of Lords, she could sense her power. I wouldn't say that it went to her head, but I was perhaps the only person who had the courage to criticise her, and she didn't like it."[93]

Three days after that article was printed, Greenfield was interviewed also by the *Daily Mail*—in an article with the headline quoting her: "Peter's Brain Was a Real Aphrodisiac, but Now I Can Wear Short Skirts without Being Nagged." And she noted that Atkins hated some of her clothes, especially platform shoes. She said they drifted apart and that she did not want to be "committed to one man, to have my freedom constrained" and that "when I thought about how busy I am with my public life, marriage to Peter no longer seemed the best way to spend any free time I had."[94]

Looking back on the coverage, she said: "I was surprised people found it that interesting, to be honest. Also, I was less experienced in those days. I very foolishly spoke or said something briefly to the *Daily Mail* and then my husband, who is even less experienced than me . . . just splurged to the press. That was upsetting, because you do feel invaded."[95]

Female critics offered different interpretations of this relentless media focus on Greenfield's personal, feminine, and sexual dimensions. For Fara, Greenfield showed that "science and sexiness are compatible for women as well as for men."[96] For Mwenya Chimba and Jenny Kitzinger, Greenfield's representation could be positive—women scientists are "fashion-conscious" instead of "frumpy"—or could be negative, leaving her open to "the implicit accusation that she is being manipulative and using her sexuality to attract attention." These authors argue, furthermore, that the idea that everyone has to be "brilliant *and* gorgeous can be seen as deeply unhelpful" to promoting women and science.[97]

Nature described the multimedia portrayal of Greenfield in London's National Portrait Gallery—a continuously shifting picture of her created from about two hundred pieces of artwork—as presenting "a scientist whose gender is irrelevant, but whose intellectual curiosity is captured in the semi-abstract, constantly changing representation."[98]

CRITICIZING DIGITAL CULTURE

Greenfield continued to advance her ideas on consciousness through popularization. *The Private Life of the Brain*, first published in 2000, argued that the mind and the emotions were part of a continuum: the greater the emotion, the lesser the consciousness and vice versa. The emotion-driven behaviors that swept away the sense of self included those of ravers high on the drug ecstasy and schizophrenics, who had feelings of terror and rage. The greater their emotion, the less their conscious action, and, consequently, their individual identity was reduced and they were left, ultimately, trapped in the present.

The Private Life of the Brain was a footnote-laden popular book that featured a more specialized scholarly appendix, an indication that Greenfield wanted to maintain her professional credibility for her expert readers. The historian of medicine Roy Porter praised her evaluation of competing theories of consciousness and the "enviable clarity" of her prose and her "assured personal narrative voice."[99]

This theory of consciousness as the personalized physical brain underpinned her subsequent popularization work. As her profile grew, her writing explored more expansive terrains. *Tomorrow's People*, first published in 2003, was a speculative description of an imagined future shaped by genetics, information technology, and nanotechnology. The intellectual and emotional lives of tomorrow's people were altered by a technology that offered the intense sensation of cyber escapism or designer drugs. The constantly wired society shaped the personalized brain, leading to collective consciousness and loss of individual identity, liberty, creativity, and originality.

Greenfield said the book "should really have been a novel" and was originally conceived as one with "a brilliant and beautiful heroine, a female neuroscientist."[100] She began to write it while on a Christmas break in the Caribbean, but abandoned it days later as she was frustrated by her weak literary efforts. The book provided a forum for her to imaginatively explore the future without needing to conform to her peer community's standards of evidence. She wrote: "As a research scientist I get to speculate from time to time, planning a new set of experiments or trying to interpret a puzzling finding. But speculation unsubstantiated by published data holds no currency."[101]

Critics attacked *Tomorrow's People*. Once the book moved from the neuroscience of the brain, the *Observer* noted, Greenfield's arguments were "infuriatingly unsubstantiated and one-sided, peppered with dangerously unqualified statements and reductions ad absurdum" that ignored "the influence of essential human nature" that her classics education might have inculcated.[102]

The criticism of *Tomorrow's People*—that its arguments are too broad and too unsupported by evidence—mirrored the arguments made against Greenfield's scientific research. A recurring public criticism of Greenfield was that, despite her prominence, she has not been an influential scientist. *Science* offered a measured summary of this criticism when it wrote that she appeared "to have left real science behind without delivering on the promise of her early ideas."[103]

But other criticism was usually caustic. And it was delivered anonymously. An unnamed scientist asked in the *Guardian* whether "she had ever really produced anything of any importance."[104] Another scientist called her work "very ordinary."[105] A scientific contemporary of hers told the *Observer* that her celebrity status allowed her to circumvent the accepted quality-control measures within science.

"With fame, she has become detached from all the processes of scrutiny and quality control that scientists use when they communicate with each other through papers or whatever," said the scientist. "A lot of what she says does not pass muster academically. Britain is very strong on neuroscience and compared to the leaders in the field, she is simply not in the same league. She is never cited in research papers."[106]

Her peers attacked also her popularization work and her presentation as a public scientist. An anonymous neuroscientist described "her absolute lack of the kind of decorum that befits a scientist in the public sphere."[107] Greenfield once told a reporter: "I wear what I wear. Criticism is predicated on a curious assumption about scientists. If I was an advertising executive, would that be an issue?"[108] The *Sunday Times* wrote that her "bitchier colleagues suggest that the reason she looks more like an advertising executive than a don is because she is one and the brand she is promoting is herself."[109] A Royal Society member complained anonymously to the *Independent*: "She doesn't promote science. She promotes Susan Greenfield."[110]

This anti-Greenfield sentiment was illustrated most sharply in the coverage surrounding the rejection in 2004 of her nomination to become a Fellow of the Royal Society (FRS), the world's oldest scientific society. She was nominated for membership by secret ballot in December 2003 but was rejected in February 2004. One scientist said anonymously that, if she were elected, Sir Isaac Newton "would turn in his grave."[111] Another anonymous source said that neither her research nor her popular writing were good enough. Granting her membership, said the same source, "would be an insult to the world-class scientists who are still on the waiting list and to the legions of modest, hard-working genuine promoters of public dialogue on science."[112]

If Greenfield became an FRS, said another anonymous source, it "would be an unjust reward for self-promotion" and several female fellows opposed her nomination.[113] Greenfield responded through her lawyers and said it was "a great pity that those who do not have the courage to identify themselves can make unsubstantiated criticisms of my science and my activities in public communication."[114]

The rejection highlighted what one journalist for the *Guardian* called the crux of the "Greenfield problem." That is, it was laudable for scientists to discuss their specialty, but confident communicators who succeed as media figures were viewed as hungry for publicity. Greenfield, the paper concluded, "has paid the price of media success."[115] The *Times* asked in a headline if she was "Too Famous to Be a Fellow?"[116]

Greenfield interpreted her career-long criticisms as gender-driven. "What have I actually done to deserve these comments?" she asked. "I suspect the detractors are male, just because of their sheer numbers in the Establishment."[117] Asked if the Sagan Effect had dissipated, she said: "Sadly I don't think so. I think that people are always suspicious of things that are different from them or that threaten them. And when those two things come together, then they are very suspicious of it."[118]

THE CONTROVERSIAL
ADVOCATE OF MIND CHANGE

Following her Royal Society rejection, Greenfield continued to propose scientific ideas in popular books. *ID: The Quest for Meaning in the 21st Century*, first published in 2008, built on the idea of the mind as the personalized brain. Greenfield used this idea as a scientific foundation for an exploration of the future identities of a generation wedded to electronic media. The changes to malleable neuronal connections caused by new interactive technologies were unknown, but she hypothesized that the constant short-termist pleasure seeking that characterized social networking and computer gaming might lead to emotional detachment and a reduction in an ability to imagine, analyze, and think in abstraction.

Greenfield postulated that the insatiable learned desire for immediate stimulation was linked to gambling, antisocial behavior, and obesity and drove demand for Ritalin, a medicine used to treat attention deficit hyperactivity disorder (ADHD). The sensation-saturated screen culture, driven by instant gratification, needed constant rapid stimulation and led to reduced empathy and sense of self. She wrote: "if the old world of the book aided and abetted the development of a 'mind,' the world of the screen, taken to extremes, might threaten that mind altogether, and with it the essence of you the individual."[119]

Critics viewed *ID* similarly to *Tomorrow's People*. The *Times* argued that the book convincingly described Alzheimer's disease, but its social commentary was a "hypothesis in search of evidence."[120] The *Mail on Sunday* said: "The book has a makeshift, undigested quality, veering to and fro between the obscure and the bleeding obvious."[121] The *Sunday Times* said *ID* "digresses all over the place in little flash floods of maddening provisos and second thoughts. It's as if she dictated it while bouncing on a trampoline, fixing an errant eyelash and sorting her fraught schedule on a BlackBerry," but yet had "more than a drift of horrible truth."[122]

But her status as a media and political figure meant Greenfield was in a position to communicate *ID*'s central warning—immersion in digital media could radically alter personalities—to various audiences, through different channels. In 2009, on the fifth anniversary of Facebook's creation, she asked two questions on this topic in the British Parliament: "First, why are social networking sites growing? Second, what features of the young mind, if any, are being threatened by them? Only when we have insights into these two issues can we devise more general safeguards, rooted not so much in regulation as in education, culture and society." Then, building on the theory of the mind as built on the plasticity of brain cells, she asked: "It might be helpful to investigate whether the near total submersion of our culture in screen technologies over the last decade might in some way be linked to the threefold increase over this period in prescriptions for methylphenidate, the drug prescribed for ADHD."[123] Her ideas, articulated in popular science, were inserted, via her unique political position, into the permanent government record.

Greenfield also authored two first-person articles about the dangers of screen culture in the *Daily Mail*, a socially conservative newspaper sympathetic to her thesis. One had the headline: "God Help Us All when Generation Text Are Running the Country."[124] The other was titled: "How Facebook Addiction Is Damaging Your Child's Brain." In this second article, she drew on her scientific credentials as a foundation for her social commentary. Greenfield wrote: "As an expert on the human brain, I am speaking out as I feel we need to protect the young."[125] She was one of 110 signatories—

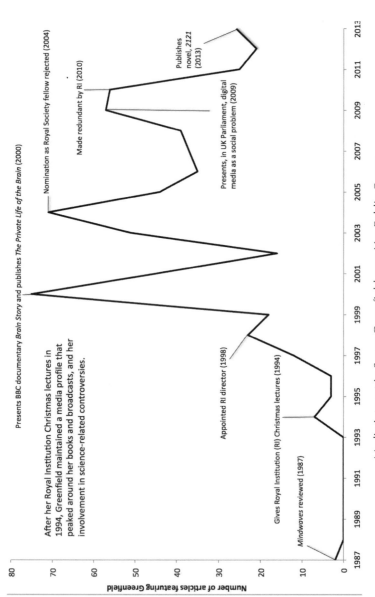

Media Interest in Susan Greenfield over Her Public Career

others included teachers, psychologists, and novelists—to a letter published in another conservative newspaper, the *Daily Telegraph*, which argued that childhood was harmed by a culture of junk food, marketing, video games, and school targets.[126]

The commentary about Greenfield's interventions around digital culture exposed different views about the role of the public scientist. One columnist said the comments were her "latest, astonishingly unscientific outburst" that was motivated in part by Greenfield's need for "recognition in order to confirm that she exists."[127] However, a writer for the *Times* supported Greenfield's stated aim of provoking debate: "there is logic in using her self-styled position as chief populariser of science to hypothesize, which she justifies as a staple of the scientific process."[128] Greenfield defended her public hypothesizing. "It's not as if I seek publicity or want it or stress too much over it," she said. "I'm asking questions rather than giving answers and that, I think, is very hard for [the media] to grasp. I'm saying I'm concerned about the rise in autism. I'm concerned about the rise in ADHD, so why aren't we talking about it?"[129]

In 2010, Greenfield was made redundant by the Royal Institution after it effectively abolished her position. (Hours later, she was reportedly locked out of her apartment.)[130] Describing the fallout, the *Guardian* in an editorial said the conflict pitted "one of Britain's most outspoken scientists—a sharp, quotable and persuasive media star—against one of the country's most venerable scientific institutions." The editorial said her "profile and manner were the reasons the RI appointed her. It is unfortunate that they now seem to be part of the reason she was removed."[131] As the chart shows, media coverage in her career initially focused on her broadcasts and books, but as her public career progressed, journalists reported on controversies in which she was involved. She was an established and controversial public figure who did not need a new book or TV show to draw journalistic attention.[132]

#GREENFIELDISM

Greenfield continued to articulate her theories in public. The hashtag #Greenfieldism entered the Twitter lexicon in 2011 after Greenfield argued in a *New Scientist* interview that constant immersion in screen technologies damaged children's developing brains and had the potential to make them more violent, distracted, and addicted. She argued that the rise in the use of new media mirrored an increase in numbers diagnosed with autistic spectrum disorders. As evidence, she cited articles in the scientific journals *PLoS One* and *Neuron* but noted that there were constraints on the collection of evidence for an emergent social problem. The effects of screen culture could only be seen in twenty years, and these effects could not be tested using experiments. But she argued: "I think there are enough pointers that we should be talking about this rather than stressing about not being able to replicate things in a laboratory instantly."[133]

Autism support groups and experts immediately contested the claims. The Oxford professor of neuropsychology Dorothy Bishop dismissed Greenfield's comments as being without evidence and more of the "illogical garbage" told to the parents of autistic children.[134] Defending herself, Greenfield told the *Guardian*: "I point to the increase in autism and I point to internet use. That's all . . . I have not said that internet use causes autism."[135]

Her comment led to a flurry of public evaluations about the nature of scientific evidence. Eminent science writer Carl Zimmer wrote on Twitter: "I point to the increase in esophageal cancer and I point to The Brady Bunch. That's all. #Greenfieldism." This was followed by a spate of other tweets that critiqued Greenfield's claim of correlation. For example, one said: "I point to the internet, and I point to the financial crisis. That's all. #Greenfieldism."

But one point she made in an interview with the *Guardian* underlined the social value of her public intellectual work. Greenfield here took on the role of a modern expert who justified her conclusions in the light of public scrutiny.[136] She told the paper: "It could be the case that this different environment is changing the brain in an unprecedented way. It's such an important issue and I'm just putting it before people to discuss. . . . All I have ever said is, let's talk about this."

You and Me: The Neuroscience of Identity (2011) built her theories of mind and consciousness to outline a theory of identity that was based on neuroscience. For Greenfield, identity was grounded in neurological explanations of mind and consciousness, and was not something absolute, having a degree of flux.[137] This definition meant that screen technologies had the potential to change the physical structures of the brain and ultimately affect a person's identity.

Greenfield's late-career writings, part science, part social commentary, can be viewed within a matrix of similarly themed texts that have pointed to the Internet as a social problem. Science studies professor Sherry Turkle argued in *Alone Together* (2011) that connection-promising technology becomes a substitute for real-life intimacy. Nicholas Carr's *The Shallows* (2011) argued that screen culture inhibited the deep thinking needed for scientific and artistic work. Reviewing the book, Greenfield wrote: "screen technologies are neither evil nor miraculous in their effects on the human mind. . . . What is certain, however, is that our minds will change."[138] Indeed, Greenfield framed the potential harm of screen technologies as a major global social problem. "My own view is that these changes in the brain could have implications that are as serious and pervasive as those for climate change," she wrote. "Annihilation of identity would almost be as bad as the annihilation of the planet. . . . The crucial issue is how future generations will think and feel: 'mind change.'"[139]

In response, Goldacre asked, "Why, in over 5 years of appearing in the media raising these grave worries, has Professor Greenfield of Oxford University never simply published the claims in an academic paper?"[140] Reflecting on this and similar criticisms, Greenfield said a large number of her approximately two hundred papers concern the neurological environment of the brain.

"To criticize someone because you haven't done a single paper on a broad concept, it's a bit like saying you should have done a single paper on climate change, which you can't do," she said in an interview. "I think if that is a criticism, it is not a very fair one, because it shows they don't really understand that's there's a difference between a broad concept like climate change—or mind change—and then a specific hypothesis that's testable with a very particular technique in a particular experimental paradigm."

She said: "What one does, if one's an intellectual, for want of a better word, is you look at the literature, you distill ideas, you try to have general themes, you try to work within a broad framework." For her, the evidence emerging about new technologies' effect on the brain constituted the early stages of a paradigm shift, in Kuhn's terms. But there is great social resistance to these conclusions. She said in an interview: "I think that this has been more resistant even than climate change would be. It's like smoking and cancer. If you have a lot of people who are enjoying doing something and other people are making a profit out of it, you are not going to on the whole have easy traction."[141]

THE SCIENTIST AS NOVELIST

Mind change was the core idea in her novel *2121*. The book, published in 2013, described a dystopian society split into hedonistic and impulsive

"Others" and cloistered and cerebral "neo-Puritans." It tells how neo-Puritan Fred goes to study the Others and falls for Zelda. To illustrate the novel's explicit thesis that the addiction to technology distorts thinking and feeling, Fred and Zelda's interactions bring both of them to more moderate, balanced personalities. Critics savaged *2121*. The *Guardian* said it was "badly conceived, badly realized, badly characterized, badly paced and above all badly written."[142] The *Observer* criticized her for presenting her scientific agenda, not as a dialogue, but in a form where she cannot be contested. "Greenfield has evolved into a figure of great controversy for persistent vocalization of her technofears," wrote the reviewer. "The characters and her dystopia are one-dimensional in order to drive Greenfield's broader technophobic narrative."[143]

However, Greenfield said the novel was not a vehicle for her ideas. Mind change was instead an intriguing theme for a novel. The novel also marked a more sustained return to public life. Before it was published she visited—on the advice of her public relations company—the *Financial Times* office for an interview. She handed the reporter a twenty-page printout of references to studies related to mind change. "The outspoken neuroscientist," the paper said, "is back in action."[144]

Mind Change (2014) developed her arguments in detail, based on a foundation of hundreds of references. It can be viewed as a discussion of science-in-the-making, a discussion only a star scientist could introduce in public. "I've often seen my role, not so much that I'm bringing down tablets from the mountaintop, but more I'm acting as a catalyst to get people to think a bit more and to challenge things and to think for themselves," she said in an interview. "By my own lights, that's what a twenty-first-century scientist should do."[145]

GREENFIELD: A THREE-ACT TALE OF SCIENTIFIC FAME

Susan Greenfield exemplified two forceful, intersecting trends in modern science: the drive to entrepreneurial science and the new value of public visibility. A commercially savvy scientist in 1980s free-market Britain, she received major funding from pharmaceutical company Squibb for her research and was praised by the business community as a modern scientific entrepreneur who had a financial—as well as intellectual—stake in the molecules she studied. Today university scientists have an ever-closer relationship with industry and the ability to raise funds is now a routine part of how a scientist's career is evaluated. Greenfield identified and surfed this post-1960s entrepreneurial trend as it broke.

She also came to embody a changed system of values in science. The traditional indicators of a scientist's worth were research prowess and peer prestige. But in today's public science marketplace, additional gauges of quality are media profile and popular appeal. These qualities granted her high status within the scientific community. She became the media-friendly modernizer who could drag the dusty Royal Institution into the new century, infusing it with her singular combination of style-magazine glamour and Oxford intellectualism.

Though she succeeded in bringing the RI into the public eye, her public life demonstrated the pitfalls of fame. She traded trivial private details for promotion, but—like Stephen Hawking—she found that media coverage cannot be controlled. Seeing the details of her marriage discussed over pages and pages of the *Daily Mail* showed how the private life of public scientists interests people (and not always for the most noble of reasons).

Her fame illustrated also the particular risks that female public scientists face. She was portrayed over and over in stereotypically feminine ways, with a focus on her looks, clothes, and relationships, a portrayal she was at least complicit in creating. Even as she wrote about the unfair burdens placed on female scientists, she was trapped within a set of media portrayals that stressed her femininity and her perceived sexual attractiveness. Greenfield was exhibit A in modern media's dubious shift to casting female scientists in roles that stress their looks and glamour.

There is an element of tragedy in the way she was ousted from the RI. She was hired in part to give it populist appeal and cultural cool. Yet her style of stardom contributed to her downfall there. If her rise to fame was the first act of her public career, and her fall was its second act, then her star status has granted her a third act. Her profile and reputation granted her the platform to raise in broad culture the uncertain, contested, and controversial idea that living online is changing our brains and personalities. Celebrity grants authority to today's public scientist.

SEVEN

The Reluctant Fame of James Lovelock

Silent Spring haunted the 1960s public imagination. The book's vision of a spring without birdsong symbolized the ominous future that the embryonic global environmental movement fought to prevent. Rachel Carson grounded her arguments in scientific data collected by a machine invented in 1957 by one of the modern era's most unorthodox scientists. James Lovelock's electron capture detector measured with acute sensitivity traces of chemicals, such as the pesticide DDT, Carson's chief chemical villain, in air and water. Lovelock later wrote that his palm-sized invention was "without doubt the midwife to the infant environment movement."[1]

Yet Lovelock, one of the world's few independent scientists, who worked for decades out of his own laboratories in two rural British farmhouses, did not become a public figure because of his world-changing invention. He imprinted himself on the public mind instead as the creator of Gaia, the idea of the Earth as a single superorganism, an idea the *Independent* called "the most radical way of looking at life on Earth since Darwin."[2]

The idea made Lovelock "one of the most famous scientists on the planet,"[3] the "godfather of modern environmentalism,"[4] and the "godhead of the ecology movement."[5] *Rolling Stone* labeled him "one of the twentieth century's most influential scientists." A climate scientist once predicted Lovelock's ideas would culminate in a reworking of science in a manner similar to Copernicus.[6]

Lovelock's public career shows how an unconventional idea once ridiculed as a form of "pseudo-scientific myth-making" can be spread through popular media, sustained in odd ways in popular culture, and eventually embraced by the scientific community and the wider public as a clear and powerful way of understanding our climate crisis. While Susan Greenfield embraced fame and its power to enhance scientific reputations, Lovelock only sought out popular media when his ideas were effectively shut out of mainstream science. His fame was a reluctant fame. But nevertheless, his fame dramatically shows how the workings of science can be affected by the power of celebrity.

FORMULATING AND FRAMING A BIG IDEA

James Lovelock trained as a chemist, and graduated from the University of Manchester in 1941 and later earned a PhD in medicine from the London School of Hygiene and Tropical Medicine. For more than twenty years, he worked with the UK's National Institute for Medical Research. While there, he demonstrated that the common cold was transmitted not through the air but by touch. He also invented an early version of the microwave oven, an invention he used to reanimate cold animals for medical research (and to heat his lunch).[7] "If he had received a specialized scientific education," the British philosopher John Gray in the *New Statesman* would later write, "he might never have developed the Gaia theory."[8]

He began life as an independent scientist in 1964, making him a rarity in a modern scientific enterprise, where almost all research has been conducted in universities, corporations, or research centers by teams of researchers. From laboratories in his rural English homes, first in Wiltshire and later in Devon, Lovelock has worked as a freelance scientist and inventor in consultancy work—often using specialized equipment he created himself—for NASA, the British government, the security services, and multinational companies such as Shell and Hewlett-Packard.

The idea for Gaia came in 1965, as Lovelock worked on one of his first commissions as a lone scientist. NASA hired him to advise on how to find life on Mars. Working at the Jet Propulsion Laboratory (JPL) in California, he wondered why the atmosphere on Mars had a different chemical composition from the atmosphere on Earth. Then the idea, he said, came to him in a "flash of enlightenment." The Red Planet was dead. But Earth teemed with life. The life forms did not adapt, as conventional wisdom assumed, to the conditions of the planet. The Earth's living matter, atmosphere, oceans, and land instead formed a complex interconnected system that has, for the past four billion years, made the conditions of Earth fit and comfortable for life.[9]

Back in his English village of Bowerchalke, during a walk in the countryside, Lovelock explained his idea to his friend and neighbor William Golding, the author of *Lord of the Flies*, who later won a Nobel Prize for literature. The writer told Lovelock: "If you intend to put forward so large an idea you must give it a proper name, and I suggest that you call it Gaia."[10] Lovelock—whose psychological profile carried out as part of an early-career job application for pharmaceutical firm Thomas Hedley concluded that he was best suited to a career in marketing—found the title perfect.[11] "In spite of my ignorance of the classics, the suitability of this choice was obvious. It was a real four-lettered word and would thus forestall the creation of barbarous acronyms, such as Biocybernetic Universal System Tendency/ Homoeostasis," he wrote, adding, "I felt also that in the days of Ancient Greece the concept itself was probably a familiar aspect of life, even if not formally expressed."[12] The expansive idea had an evocative title.

Lovelock tried to first introduce Gaia through formal scientific channels. It was then called the Gaia hypothesis: for scientists, a hypothesis is an untested idea proposed to explain facts about the world, while a theory has been tested and considered true. He worked with Lynn Margulis, a biologist he met at the end of the 1960s who specialized in the evolution of cells and a researcher also interested in unconventional theories who had within science a "quasi-outsider status."[13] Gaia seemed to challenge established ideas about the large-scale forces that shaped Earth and the planet's life. Geologists believed geochemistry explained the workings of the Earth's crust and oceans. Biologists used natural selection to explain the development of life. Consequently, Lovelock and Margulis found it difficult to attract mainstream scientific interest in Gaia.

Nevertheless, the hypothesis was first published as a one-page statement in *Atmospheric Environment* in 1972. A longer paper was rejected by *Science* and other journals before it was published in *Icarus*, edited by Margulis's former husband, Carl Sagan. Another paper was later published in a Swedish environmental science journal *Tellus*. Neither journal was at the center of establishment science. Neither of the papers gained significant scientific traction. Nor did the papers impact significantly on public opinion.[14] Gaia looked like a scientific nonstarter.

BECOMING A MAVERICK SCIENTIST

Lovelock had become a "maverick scientist," a term used in a technical sense by scholars of science. Maverick scientists are those who champion unorthodox ideas and who are oppressed by a scientific establishment that denies them research funds, rejects their papers for publication, and shunts them to the margins of their field. Lovelock would later call himself a "scientific maverick"[15] and commentators repeatedly labeled him "a British maverick"[16] and a "maverick research chemist."[17] After he failed to persuade the mainstream scientific community to take an interest in Gaia, he followed a path often taken by mavericks: he turned to popular media.[18]

Lovelock's coauthored 1975 article on "The Quest for Gaia" made the cover of *New Scientist*.[19] The respected British weekly science magazine, aimed at scientists and nonspecialists interested in science and technology, often published hypotheses and speculations as well as established theories.[20] Its striking sci-fi-like cover on February 6, 1975, showed a man standing in a scrapheap of machinery, with an industrial plant on the horizon, gazing up at a globe of the earth. "Gaia is still a hypothesis," wrote Lovelock and his coauthor, Sidney Epton of Shell. "The facts and speculations in this article and others that we have assembled corroborate but do not prove her existence but, like all useful theories right or wrong, Gaia suggests new questions which may throw light on old ones."[21]

New Scientist jump-started Lovelock's celebrification. In the same edition, the magazine profiled him, photographing him wearing a casual wool sweater, standing in a rural landscape.[22] The reporter traced Lovelock's schooldays and his career, described his home environment, where his three certificates of recognition from NASA are mounted on the wall, outlined the working arrangements of an independent scientist: his laboratory was a converted garage and his office was in the children's former bedroom. The profile called him "one of the last of the old-style natural philosophers."[23] *Newsweek* a month later said Lovelock had "a reputation as an old-fashioned natural philosopher."[24]

CREATING A NEW LOOK AT LIFE—
THROUGH POPULAR SCIENCE

After *New Scientist* brought Gaia to wider public attention, more than twenty publishers courted Lovelock. Big ideas made scientists skeptical, but editors quickly saw Gaia as a novel concept that could be packaged, marketed, and sold.[25] As a maverick scientist, he had to choose carefully; the reputation of the publisher mattered because it would convey legitimacy on Gaia. He chose one with impeccable scholarly credentials: Oxford University Press. Moreover, the format of the book as a popular science title allowed Lovelock more freedom to infuse the book with more ideas than would be acceptable within the confines of a specialist academic monograph.

He wrote the book in a cottage near Bantry Bay on the southwest coast of Ireland. It was, he wrote, "like moving into a house run by Gaia,"[26] and the rugged, unspoiled landscape inspired the book's romantic and spiritual imagery. "I used to sit on my favourite slab of rock overlooking Bantry Bay and the broad Atlantic," he later wrote. "As I sat in the warm sun on my ledge, high up on the sandstone slabs of Hungry Hill, it was not easy to think about the Earth in any way except romantically."[27]

Gaia: A New Look at Life on Earth, first published in 1979, mixed science, spirituality, and ethics. Lovelock outlined in detail the concept of the Earth as a sort of living organism. Lovelock saw *Gaia* as more than a scientific treatise in popular form; he said he did not write it just for scientists, but also engineers, physicians, practical environmentalists, and nonspecialists who needed "moral guidance" in their work. He noted that he did not conceive of Gaia as sentient, but it was almost impossible to write without referring to Gaia as alive—he said it was like referring to a ship as "she." "The entire surface of the Earth including life is a superorganism," Lovelock wrote, defining the concept, "and this is what I mean by Gaia."[28]

Nature liked *Gaia*. The journal's reviewer read it "with immense pleasure," welcoming its novel perspective of the Earth as a superorganism, but warning that its ideas would likely require future scientific refinement.[29] Despite his maverick status, Lovelock's pre-Gaia reputation granted him some legitimacy. "A surprising idea from an established figure," wrote one science studies scholar, "will be seriously considered."[30] And Lovelock had been a Fellow of the Royal Society since 1974.

But evolutionary biologists savaged the book. By the late 1970s, they had become the scientific establishment, the influential gatekeepers who could dismiss as pseudo-science ideas that did not conform to their view of science. And they viewed Gaia as a threat to their foundational and fundamental theory: Darwinism.[31] A review in *Co-Evolution Quarterly* said the book presented an "unquestionably false" view of natural selection.[32]

The idea of a living planet, where all life forms somehow evolved together for the good of all species, cut against natural selection, particularly the view that life forms were driven fundamentally to ensure their own survival. Richard Dawkins characterized *Gaia* as "pop-ecology literature." If Gaia theory was compatible with evolutionary biology, he argued in *The Extended Phenotype* (1982), there would have to be a process of interplanetary natural selection. The universe would be full of dead planets unable to sustain life, but it would also contain planets, such as the Earth, that successfully adapted to regulate life.[33] Stephen Jay Gould called Gaia "warm and fuzzy," "a pretty metaphor, and not much more."[34]

Gaia had first been shunted to the edge of science. Now it was shot down in popular science.

Dismissed hypotheses usually die. But Gaia continued to live because it connected with religious and spiritual communities. "To my astonishment, the main interest in Gaia came from the general public, from philosophers and from the religious," wrote Lovelock. "Only a third of the letters were from scientists."[35] The late Bishop of Birmingham, Hugh Montefiore, saw Gaia as religious. "It made me think of what theologians call the immanence of God," he said. "That is to say, the Holy Spirit of God working within creation."[36]

The New Age spirituality that was a feature of late-1970s counterculture in the United States, particularly, worshipped Gaia as a sort of living Earth mother. Gaia Books was founded in 1982, publishing texts on science, environmentalism, and spirituality, the three elements of "the Gaian life-style."[37] The same year, Fritjof Capra in *The Turning Point* (1982) championed the rise of a new framework for grand unified thinking, into which Gaia moved seamlessly. "No wonder the professional scientists were horrified," wrote the philosopher of science Michael Ruse. "It is often not the content but the company it keeps that renders a concept pseudo-scientific."[38]

By the mid-1980s, Gaia seeped into popular culture in other ways. The BBC's acclaimed drama about the murky politics of nuclear energy, *Edge of Darkness* (1985), featured an environmental terrorist group called Gaia. Science fiction writer Isaac Asimov published *Foundation and Earth* (1986), a novel about a planet with one mind called Gaia. Lovelock's coauthored novel, *The Greening of Mars* (1985), told how humans created an earthlike atmosphere on Mars by altering its atmospheric chemistry. Carl Sagan reviewed it favorably for the *New York Times*: he called it science fact alluringly presented as science fiction.[39] In an article alongside Sagan's review,

Lovelock admitted he used Mars as a theater to showcase his Gaia theory. "Fiction," he said, "always seems more credible than fact."[40]

Amid these portrayals, the *New York Times* published a lengthy profile of Lovelock in 1986, presenting him and his work as an odd juxtaposition between high-tech science and unspoiled nature. "Not far from a river in St. Giles-on-the-Heath, a hamlet on the Cornwall-Devon border in southwestern England, stands the laboratory of James E. Lovelock, a white, windowed cabin attached to his house and patrolled outside by half a dozen peacocks," wrote science writer Lawrence E. Joseph. "A phalanx of spectroscopes, radiation detectors and microcomputers set incongruously in a fragrant meadow, the lab seems like a probe sent to unravel the secrets of nature."[41]

THE EMBLEM OF ECOLOGY

These juxtaposed notions of technology and nature, reason and romanticism, science and spirituality all featured in Lovelock's public image. All these elements were part of ecology, the scientific discipline that came to enhanced prominence in the late 1960s and 1970s as environmental consciousness developed. At the time, one of the discipline's most famous scientists, Barry Commoner, described his four laws of ecology: "everything is connected to everything else, everything must go somewhere, nature knows best, and there is no such thing as a free lunch."[42]

But nonscientists have not always viewed ecology as a science. They saw it as an ethical standpoint, a guide to how they could live harmoniously with an environment they revered. Or they saw it as a political movement, a counterpoint to what they considered the capitalist exploitation of nature. Or they saw ecology as a big-picture framework that made the seemingly separate parts of the world coherent and connected.[43]

Moreover, ecology, like environmentalism, was ambiguous about *science itself*. Science was seen as the cause of ecological harm, but also the only way to rectify that damage. For some environmentalists, science was the source of their authority. For others, science lacked authority because it was tied to industry and the military.

Lovelock and Gaia symbolized these issues—in different ways for various citizens. Lovelock embodied the core characteristics of the ecologist. As the intellectual historian Anna Bramwell argued in *Ecology in the*

20th Century, ecologists believed in a harmonious nature, engaged with the natural world, adhered to aesthetic values that were close to spiritual values, opposed the idea that humans and nature were separate, believed in objectivity, yet had some skepticism about what they termed traditional science. The application of ecological ideas to politics and culture, a form of criticism called political ecology, featured ethical and moral claims, dramatic remedies, and apocalyptic scenarios.

The scientific community looked intently at Gaia again in 1988. The American Geophysical Union conference, a respected venue where experts came together each year to discuss the latest developments in their field, zeroed in on a weakness in Gaia—it was difficult to test. A hypothesis that could not be tested failed a fundamental scientific rule: it could not be proved false and, therefore, was scientifically worthless. The conference brought together specialists from diverse fields to create practical and workable tests to Gaia. A *Nature* report of the meeting said the idea "seems to have had a fair hearing."[44] *Science* in its report said Gaia had become "respectable."[45] The meeting led to publication of an account of the papers presented at the conference, *Scientists on Gaia* (1991).

But Lovelock regarded the conference as a disaster. He believed that the most persuasive voices at the conference were the ones that dismissed the idea as difficult to define and almost impossible to test. The scientific community looked hard at Gaia. But it remained unmoved.

One scientist had enough of Gaia. A month after the conference, well-known British microbiologist John Postgate used a *New Scientist* article to attack Gaia and blame journalists for keeping the discredited idea on the public agenda. "Am I the only biologist to suffer a nasty twitch, a feeling of unreality," he asked, "when the media invite me yet again to take her seriously?" The lack of engagement with Gaia from researchers was "the silence of politeness" caused in part by Lovelock's earlier scientific achievements.

Postgate felt he had to break this silence because Gaia was "silly and dangerous," a form of "pseudo-scientific myth-making" akin to the fads that had surged in prominence as science fell into disrepute since the 1960s: "astrology, fringe medicine, faith healing, nutritional eccentricities, religious mysticism and a thousand other fads and cults which now plague western societies." Postgate worried about a future where "hordes of militant Gaiaist activitists [enforce] some pseudoscientific idiocy on the community, crying 'There is no God but Gaia and Lovelock is her prophet.'"[46]

MOVING GAIA FROM
PUBLIC LIFE INTO SCIENTIFIC LIFE

Nevertheless, Lovelock persisted. *The Ages of Gaia*, first published in 1988, was a substantially more technical book than *Gaia* and sought to establish the idea's scientific credentials. The book rebranded Gaia as a new science that combined earth sciences and life sciences—geophysiology. It also eradicated an error that scientists had long banished from their ways of thinking, an error that Lovelock's critics argued he made again and again: the scientific sin of teleology: the argument that the Earth and its life forms had motive, purpose, and destiny.

A chief aim of the book was to counter the devastating critiques of evolutionary biologists. The way to do this was to describe how Gaia evolved. Lovelock's solution was to present as evidence for evolution a computer model he designed. The model described the workings of a pretend planet called Daisyworld, an Earth-like planet spinning around a sun and populated entirely by different types of daisies whose growth and death showed how a stable tempererature for life was maintained on the planet.[47] *The Ages of Gaia* was the book Lovelock wanted his peers to read. Peers vetted the manuscript, he said, "as thoroughly as a paper in a science journal."[48] One reviewer made about 250 comments.

Although he presented a stripped-down scientific Gaia, Lovelock included spiritual reflections in *The Ages of Gaia*. He named one chapter "God and Gaia," and argued that his idea was "a religious as well as a scientific concept, and in both spheres it is manageable," that "God and Gaia, theology and science, even physics and biology are not separate but a single way of thought," and that the "life of a scientist, who is a natural philosopher, can be devout."[49] Lovelock said he was agnostic and wrote: "In no way do I see Gaia as a sentient being, a surrogate God."[50]

Reviewing the book, the late science journalist Dennis Flanagan—the editor of *Scientific American* for almost four decades—identified the fissure that ran across Lovelock's idea and public persona. Gaia was, he wrote, a scientific concept trapped in the "fugue between facts and big ideas" and Lovelock was "a man torn." Lovelock's "big idea is stated with much grandiloquence and little qualification, but his arguments in favor of it are modest, qualified, plausible and above all illuminating," wrote Flanagan in the *New York Times*. "Although his own approach is not mystical, he appears to be sympathetic to such people because they share his reverential attitude toward the earth as a living system."[51]

Even so, he noted that Lovelock was no longer an outcast in scientific circles, as the Gaia hypothesis had suggested productive avenues of research. By 1988 Lovelock's ideas touched on another idea that began to grip scientific and political interest: global warming.

That year James Hansen of NASA warned a Congressional committee that global warming was real. He told elected officials he was 99 percent certain that rising global temperatures were due to the greenhouse effect, the warming of the Earth's climate caused by increased amounts of carbon dioxide and other gases released into the atmosphere. By burning fossil fuels and other activities, humans had changed the global climate for centuries to come.

Now global warming was political. In 1988 also, British prime minister Margaret Thatcher—like Lovelock, a trained chemist—became the first major world leader to identify climate change as a key challenge facing civilization. Lovelock was one of a small group of scientists invited in 1989 to give one of the first briefings on climate change to Thatcher's cabinet at 10 Downing Street. The environmental movement in the 1980s took up climate change as its main cause.[52]

As global warming worked its way onto political agendas in the early 1990s, interest in Gaia increased. The journalist who profiled Lovelock for the *New York Times*, Lawrence E. Joseph, published *Gaia: The Growth of an Idea* (1991). Joseph's sympathetic account praised how Gaia prompted new thinking in science, economics, altruism, and justice. The early 1990s also saw a spate of books on the religious dimensions of ecology, including *The Way: An Ecological World View* by the founder of the influential *Ecologist* magazine, Edward Goldsmith. Ecology as faith, wrote one reviewer, was "a refuge from scientism, mindless consumerism, and churchy religion."[53]

Lovelock developed his ideas. He released *Gaia: The Practical Side of Planetary Medicine* (1991), a text published by Gaia Books that introduced for a wide audience another metaphor: the planet as sick patient. Scientists' understanding of the planet's ecology, he argued, was analogous to doctors' conception of the body before twentieth-century medicine. And he diagnosed the planet's pathogen: humans who ate its resources and overloaded it with waste. Earth suffered a plague of humans.

The book's publication brought more public exposure for Lovelock. A journalist for the *Independent* visited Coombe Mill, where the scientist lived with his second wife, Sandy. The writer noted that for years Gaia brought pilgrims to Lovelock's door, "like the autograph hunters who tracked Charles Darwin down in rural Kent after the publication of *The Origin of Species*." The reporter called it a "technological Arcadia" and noted how Sandy organized "the huge inflow of Gaiana: scientific papers, press cuttings, letters from fans, from religious maniacs, and from people who say that her husband's books have changed their lives, a proportion of whom—to the Lovelocks' dismay—turn up to pay their respects in person."[54]

Lovelock reflected on how he has been represented: "The last journalist who came down here said something about me being a shock-haired 110-year-old mad scientist living in an enchanted wood with a nubile blonde wife."[55]

The scientist's status increased in the early 1990s. He welcomed the global press to his house, and toured the world giving speeches and attending conferences. He became so exposed in media, wrote environmental journalist Fred Pearce, a keen follower of Lovelock's career, that the scientist was no longer on the fringes of popular culture as he was in the 1970s. "Now you can read about Lovelock in the color supplements; Margaret Thatcher sought his advice; [Green advocate] Jonathon Porritt nominated him as his hero; and fashionable presenters interview him on TV," wrote Pearce. "It's good for business, good for science and probably good for the planet, but I am beginning to yearn for the return of the mad boffin of Bodmin Moor."[56]

Pearce also argued against any rebranding of Gaia. For him, the value of the theory lay in the way it lent itself to various interpretations by different groups, making it meaningful in different ways for environmentalists and young scientists. "Gaia as metaphor; Gaia as a catalyst for scientific enquiry; Gaia as literal truth; Gaia as Earth Goddess. Whoever she is, let's keep her," wrote Pearce. "If science cannot find room for the grand vision, if Gaia dare not speak her name in Nature, then shame on science. To recant now would be a terrible thing, Jim. Don't do it."[57]

Lovelock's position as a notable member of British public life was further cemented when London's National Portrait Gallery added a 1993 photograph of him to its permanent collection. Photographer Nick Sinclair photographed Lovelock at the scientist's home in Devon, standing in the thirty acres of woodland he planted decades earlier. In the shot, Lovelock clasps his hands, signifying spirituality. His white hair and glasses suggest wisdom. The natural background stands as an emblem of scientific ideas. Visually, Lovelock merges with Gaia.

James Lovelock photographed in 1993 at his former home in rural England, standing in acres of woodland he planted decades earlier. The picture is part of the permanent collection at London's National Portrait Gallery, making it an authoritative portrayal of the father of Gaia. Nick Sinclair, National Portrait Gallery collection

CLIMATE CHANGE PERSONIFIED

At the turn of the twenty-first century, scientists treated core Gaian ideas—about the Earth as a complex, interacting system that regulates itself—as vital ways of understanding and tackling the imminent threat of global warming. A new scientific discipline, Earth System Science, examined, in the words of *Science*, the planet as "a complex, interacting system."[58] And several organizations that examine global change produced the 2001 Amsterdam

Declaration on Earth System Science, which said: "The Earth System behaves as a single, self-regulating system."[59] The scientific establishment had seemingly adopted Gaia.

Other intellectuals saw Gaia as more than a scientific idea. It was a blueprint for a new way to organize society and to view human life. Philosopher of science Mary Midgley viewed Gaia as a doctrine for a secular society, a counter to the dominant individualistic ethos of Western life, a counter to what she saw as the individualism at the heart of sociobiology and neo-Darwinism. Viewing climate change from a Gaian perspective could lead to a more collectivist ethical worldview, one that could act as an alternative to atomism and individualism in society.[60]

Physicist Freeman Dyson argued that the collectivist ethos of Gaia was a welcome counterpoint to the overwhelming emphasis on competition that was ultimately harmful to the planet and society.[61] Political philosopher John Gray saw in Gaia a way to rethink the place of humans on Earth. Gaia restores a connection between humans and nature. "For Gaia," wrote Gray in *Straw Dogs*, "human life has no more meaning than slime mould." Gray argued that critics of Gaia rejected it because it was not scientific. "The truth is that they fear and hate it because it means that humans can never be other than straw dogs," revered offerings to the gods made during ancient Chinese rituals, but discarded and trampled once the ritual ended.[62]

Lovelock wrote his autobiography to help establish his place in the history of science. *Homage to Gaia*'s back-cover blurb called him "truly one of the most outstanding and creative thinkers of the twentieth century," and stated that "his work has led to the founding of the Green Movement and his famous Gaia theory has changed the way we think about the Earth." Lovelock chronicled Gaia's setbacks and successes in parallel to the peaks and depressions of his own life. During the 1980s, for example, he had terrible health problems—including a series of operations to repair a damaged urethra—and his first wife, Helen, moved into the terminal stages of multiple sclerosis. During the decade, he wrote, researchers treated Gaia "more as science fiction than science."[63] The 1990s, by contrast, saw him find domestic contentment with Sandy, and his work brought him eight honorary degrees, three major international environmental prizes, and a Commander of the Order of the British Empire (CBE), a recognition of his achievements in public life. The natural history of the Gaia concept mirrored the life story of James Lovelock.

Reviews were generally favorable and respectful. Eminent environmental scientist Stephen Schneider called the autobiography "more a baring of the soul of the man than a physiological guide to the Earth muse . . . a deeply personal self-history."[64] Schneider noted in his *Science* review that Lovelock's scientific accomplishments—about Gaia as system and the proposed mechanisms that allowed Gaia to self-regulate—were laudable accomplishments for a research team, never mind a lone scientist.

But one historian of biology sought to dispel what he considered myths about Lovelock. Adolfo Olea-Franco noted in the *Journal of the History of Biology* that researchers in their autobiographies etched an aggrandized image of themselves into the history of science. He argued that the book could lead readers to the mistaken view that Lovelock was a radical scholar. On the contrary, "most of his work as a very talented inventor of many analytical devices," wrote Olea-Franco, "was done since the middle sixties under contract with corporations (Shell, Hewlett Packard), the military (the Jet Propulsion Laboratory, NASA, the Pentagon), and even intelligence agencies such as the CIA."

For someone portrayed as a free thinker, Lovelock, the reviewer noted, had benefactors and supporters in the establishment—including Margaret Thatcher and investment banker Lord Rothschild. He also stressed other views not typically associated with radical figures: he hated Greenpeace and believed oil and chemical companies do their best to safeguard the environment.

The choice of book title was a form of self-glorification, the historian argued, because Lovelock had become so entwined with his theory in the public mind that to "pay homage to Gaia is certainly to pay homage to Lovelock."[65]

VISUALIZING THE CLIMATE APOCALYPSE

On May 24, 2004, Lovelock shattered what one journalist called "the great green taboo": he advocated for nuclear power.[66] Climate change posed such an urgent threat, he wrote in the *Independent*, that the only realistic alternative to fossil fuels was nuclear power. "Opposition to nuclear energy is based on irrational fear fed by Hollywood-style fiction, the Green lobbies and the media," he wrote. "But I am a Green and I entreat my friends in the movement to drop their wrongheaded objection to nuclear energy."[67] Although his views were not new—*Gaia* and *The Ages of Gaia* supported nuclear power as an energy source to stop the release of carbon dioxide—his article drew criticism from Friends of the Earth and Greenpeace.[68]

It was not the first time he discussed his ambivalent relationship with the environmental movement. In *Homage to Gaia* he said too many Greens were not only innocent of science—they hated it. He labeled swaths of the environmental movement as "a confused and babbling community of Green politicians and philosophers."[69] His scientific credentials separated him from many environmentalists whose ideas he equated with antiscience. "Their hearts are very much in the right place," he told one journalist, "but they often get the science wrong, and you can't really be a Green without being involved with science."[70]

Lovelock was alarmed by the latest evidence in the 2001 Intergovernmental Panel on Climate Change (IPCC) report. The world was getting warmer, the report frankly stated, and greenhouse gases were responsible. Sea levels would rise by up to half a meter, but a minority of scientists predicted a rise of a meter, leading to potential major floods by the end of the century in low-lying coastal areas such as Florida and Bangladesh.[71]

His fears heightened after he visited a British climate change research center where researchers presented evidence that showed the North Pole's floating ice was melting, Greenland's glaciers were disappearing, and tropical forests were vanishing—a series of parallel trends that dismayed Lovelock all the more because the researchers seemed to view each trend in isolation. Decades of immersion in Gaia forced Lovelock to see the whole picture. His alarm was so great that soon after the visit he started work on *The Revenge of Gaia* (2006).

The book portrayed an Earth laid waste by climate change. Worldwide temperatures would skyrocket. Ice caps would melt and would no longer cool the Earth by reflecting sunlight back into space. Carbon dioxide trapped beneath frozen tundra would escape. Sea levels would rise dramatically. Life in the oceans would die. Bangladesh would drown. India would run out of water. Billions would die. Bands of regional warlords would fight over scarce water and energy. Civilization will move north. By 2100 humans will flee to the Antarctic where the climate will resemble today's Florida.[72]

Lovelock suggested ways to absorb the climate disaster. Nuclear power must expand rapidly. Nuclear fission would supply energy until science develops technologies of safer and renewable nuclear fusion. Nuclear waste should be dumped in tropical forests where it would guard the natural habitats. Sustainable development was not a viable strategy. The plan should be sustainable retreat. The human race should write a great book, bound in leather, that would preserve the record of scientific advances that would help our descendants survive.

When the book was published, Lovelock again argued for nuclear power in first-person articles in the *Independent*[73] and the *Sun*.[74] Readers in

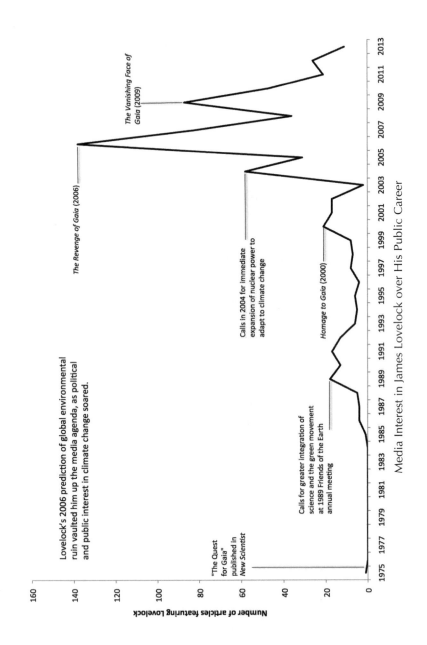

Lovelock's 2006 prediction of global environmental ruin vaulted him up the media agenda, as political and public interest in climate change soared.

The Revenge of Gaia (2006)

Calls in 2004 for immediate expansion of nuclear power to adapt to climate change

The Vanishing Face of Gaia (2009)

Homage to Gaia (2000)

Calls for greater integration of science and the green movement at 1989 Friends of the Earth annual meeting

"The Quest for Gaia" published in *New Scientist*

160   140   120   100   80   60   40   20   0

Number of articles featuring Lovelock

1975 1977 1979 1981 1983 1985 1987 1989 1991 1993 1995 1997 1999 2001 2003 2005 2007 2009 2011 2013

Media Interest in James Lovelock over His Public Career

all demographics heard his stark message. His statements had political impact, as they helped reframe Britain's nuclear energy debate.[75] As the chart shows, the publication of the books and its stark arguments led to a spike in media coverage of Lovelock.[76]

His views caught public attention as 2006 saw public and political interest in climate change spike. *Time* magazine in a cover story warned: "Be worried. Be *very* worried."[77] As the climatologist Tim Flannery would later write in the *New York Review of Books*, "The Gaia concept and climate change science are intimately connected."[78]

Against this background, environmental experts welcomed *The Revenge of Gaia*. The prominent U.S. environmentalist Bill McKibben praised Lovelock in the *New York Review of Books* as one of science's most interesting thinkers. He noted that his predictions of rapid temperature rises were bleaker than scientists publishing peer-reviewed climate research, his forecasts gloomier "than any other competent observer I am aware of."[79] Lovelock's insight was valuable because, wrote McKibben, "there are very few people on earth—maybe none—with the same kind of intuitive feel for how it behaves as a whole." Climate scientist Tim Flannery in the *Washington Post* called the book "a wondrous and novel essay, both for what it has to say and for the insight it affords into one of the most ingenious, if eccentric, minds of our age." *The Post* called Gaia "Climate Change Personified."[80]

The Revenge of Gaia's end-of-the-world vision was bleak. But for renowned environmental scientist Mike Hulme it continued a long tradition in writings about the environment—depictions of the Apocalypse. End-of-days narratives can be traced back to *Silent Spring* and they continued with the rise of global warming. "Rather than being a function of God's displeasure," wrote Hulme, "extreme climatic behavior can now be interpreted as Gaia's revenge on a morally wayward and abrasive humanity."[81] Journalists who interviewed Lovelock at this time incorporated this theme into their accounts. *Rolling Stone*, for example, called him "The Prophet of Climate Change," and noted that he was not gloomy despite his predictions that humanity faced the coming of "the Four Horsemen—war, famine, pestilence and death."[82] A *Guardian* journalist said meeting Lovelock felt "a little like an audience with a prophet."[83]

The book also drew on dystopian science fiction. The cover of the Penguin edition showed a sci-fi image of a couple locked in an embrace surrounded by swirling clouds and crashing waves as a city collapsed around them. Newspapers said the book presented a "depressing Blade Runner scenario,"[84] a "Mad Max–style vision of the coming century,"[85] and read like "the plot of one of those lurid 1970s sci-fi movies in which the world ends in an onslaught of plagues, nuclear war or apes."[86] (A reviewer would later note that Lovelock as an inventor and creator of big ideas "resembles a character out of H. G. Wells.")[87] Lovelock himself contributed to this dystopian image-association, telling an interviewer that he was not "Dr Strangelovelock."[88]

But the scientist, argued the *Sunday Telegraph*, was aware of the power of his images. The paper expressed an idea long identified by science studies scholars: that citizens assimilate fictional portrayals to form images of science. "Lovelock long ago grasped something," the paper wrote, that "other scientists are only just beginning to: that the public are more likely to be influenced by the films they see and the books they read as by dry, scientific pronouncements they only half understand."[89] Lovelock told *Newsweek*: "A cheerfully told story of doom will always prove pretty popular."[90]

A slew of books in 2006 predicted similar environmental annihilation. Fred Pearce's *The Last Generation*, for example, argued that humanity was close to a series of climate change thresholds that will lead to a global catastrophe that will kill billions, a book that science writer John Gribbin called "the most frightening book that I have ever read."[91] George Monbiot's *Heat: How to Stop the Planet Burning* argued that it was the last chance for this generation to take action. The Oscar-winning *An Inconvenient Truth* was also released in 2006 and Al Gore warned the world was in the midst of an unprecedented crisis. Science journalist Robin McKie noted that the point had come where writers on green issues had "changed their droning monologues into indignant shrieks of outrage and doom-mongering."[92]

The same dystopian themes appeared in movies and novels. The literary critic James Wood put *The Revenge of Gaia* with its "hideous picture of the warmed world" alongside postapocalyptic works such as *The Day After Tomorrow* (2004), a film that arrestingly depicts a future ice age ushered in by abrupt climate change, and *The Road* (2006), a novel that describes a future America shrouded in a nuclear winter, its scorched landscapes strewn with ash and corpses and burned-out buildings. "The possibility," noted Wood, "that familiar, habitual existence might be so disrupted within the next hundred years that crops will fail, warm places will turn into deserts, and species will become extinct—that areas of the earth may become uninhabitable—holds and horrifies the contemporary imagination."[93]

Despite the surge in media interest, *The Revenge of Gaia* failed to have the same catalytic effect on political and public life as *Silent Spring*. Lovelock quickly wrote another book, which one commentator wrote seemed "to arise from frustration that society hasn't been roused by the wake-up call of *Revenge.*"[94]

The Vanishing Face of Gaia was darker than *The Revenge of Gaia.* An abrupt temperature rise in the coming years or decades will transform the Earth, Lovelock argued, into a world of sunken cities, mass migrations, and genocidal famines. The situation, he argued, was far more grave than the IPCC predicted: its computer models did not include the sudden temperature jumps that were likely to occur as temperatures warmed. Climatologists are aware of such jumps but cannot factor them into their models because the timing and force of the leaps are so unpredictable. Yet Lovelock said he incorporated the predicted jumps into his own models.

He offered readers a reason to believe the pessimistic predictions of a "lone scientist" rather than the milder consensus of the IPCC. "I work independently and I am not accountable to some human agency—a religion, political party, commercial or government agency," wrote Lovelock. "Independence allows me to consider the health of the Earth without the constraint that the welfare of humankind comes first."[95]

Scientific reviewers were cautious. Tim Flannery in the *New York Review of Books* argued the model was not strongly predictive—essential for good scientific models—and Lovelock's entire approach needed considerable scrutiny before it is accepted.[96] Lovelock's former doctoral student, climatologist Andrew Watson, said his former mentor's expertise falters when he anticipates human actions in a shattered society. "His only special qualification," wrote Watson, "for discussing human behaviour is his longevity—having lived through the Second World War, he knows what people sometimes do to one another during evil times."

Yet Watson said although Lovelock's scientific predictions of abrupt climate change were not officially sanctioned "IPCC-speak," his warnings were nevertheless consistent with scientific consensus—and, compared with the IPCC, Lovelock's "popular voice will carry further."[97]

BUTTRESSING A SCIENTIFIC LEGACY

"Vanishing" was published alongside an authorized biography, *He Knew He Was Right: The Irrepressible Life of James Lovelock and Gaia.* Written by prolific science writers John and Mary Gribbin, the book melded Lovelock to his theory as it interwove the story of his life with the wider story of systemic scientific approaches to understanding the Earth's climate. The book presented Lovelock as a heterodox scientist, at odds with the traditional ways of doing science, at odds with mainstream science that spent decades catching up to him, at odds with blinkered thinking that carved knowledge into separate spheres.[98]

The visual images in the book reflected the myriad meanings of Lovelock. He was photographed in his home laboratory, aboard the ships *Shackleton* and *Meteor* collecting air samples, hiking on a coastal path, standing before a statue of Gaia at Coombe Mill, visiting a French nuclear reprocessing plant. Lovelock the inventor, the individualist, the ecologist, the scientist. Lovelock the maverick.

The book also aimed to cement a particular public narrative about Lovelock. "It is the stuff of legend," wrote former *New Scientist* editor Roger Highfield. "The story of the wild-eyed maverick who was attacked, vindicated and then hailed as a green visionary who could save the world."[99]

The book presented an updated argument for Lovelock's esteemed place in scientific history, beginning with the full citation to his 2006 award of the Geological Society's Wollaston Medal, geology's most prestigious honor, once awarded to Darwin. Journalists reaffirmed this narrative, describing how science welcomed Lovelock and Gaia back after he was consigned to the edge of the establishment.

Yet he continued to criticize mainstream science. After climate researchers' e-mails at the University of East Anglia were hacked and released into the public domain, there were accusations from critics of climate science and journalists that researchers altered data to better make the case for climate change. After the scandal, Lovelock told the *Guardian* he was appalled by the actions of some scientists. "What I like about skeptics is that in good science you need critics that make you think: 'Crumbs, have I made a mistake here?'" he said. "If you don't have that continuously, you really are up the creek."[100]

And Lovelock publicly corrected his own errors. In 2012, he told NBC that his climate predictions were alarmist. He said climate change is still happening, but not in the way he once thought and not as rapidly as he once feared. But he contended that *An Inconvenient Truth* had a similar degree of scaremongering. An independent scientist could admit to a mistake, but a university or government scientist could not as they feared their funding would be cut. "The problem is we don't know what the climate is doing. We thought we knew 20 years ago. That led to some alarmist books—mine included—because it looked clear-cut, but it hasn't happened."[101]

There has been an attempt to co-opt Lovelock's views into the movement variously labeled climate change deniers or climate change skeptics.[102] Asked by NBC if he was a skeptic, he said: "It depends what you mean by a skeptic. I'm not a denier."

On October 26, 2013, *New Scientist* published an article headlined "Gaia: The Verdict Is. . . ." Earth scientist Toby Tyrrell, who reviewed all the scientific studies published on the concept, concluded the planet was less stable, less robust, and more fragile than the resilient one Gaia envisaged. "My research," he wrote in the magazine, "led to a clear outcome: that the Gaia hypothesis is not an accurate picture of how our world works."[103] *New Scientist*—which introduced Gaia to the wider world in 1975—published an editorial in the same issue headlined: "Death of a Beautiful Idea." It concluded: "Gaia doesn't hold up. . . . That's science. . . . There will be no tragedy in its passing."[104]

In *On Gaia*, Tyrrell laid out the results of his tests on Lovelock's central ideas. *Nature Climate Change* reviewed it under the headline, "Requiem for a Grand Theory." Climate change expert William H. Schlesinger found the book systematic, dispassionate, and persuasive. "With the appearance of this volume, I think we can close this chapter in the history of planetary ecology," he concluded. "Gaia Is Dead; It Is Time to Move On."[105]

But Lovelock continued to recommend ways of living in a warmed world. In *A Rough Ride to the Future* (2014), he scaled down his pessimism. "I think we may muddle through," he wrote, "into a strange but still viable world." Only a few million people would survive the potential collapse of civilization wrought by climate change. The ones who live will have retreated to air-conditioned cities purpose-built for survival.

His vision of this climate-proof city is Singapore, where its 5.6 million inhabitants live in an ordered and prosperous metropolis, kept cool indoors in a hot and humid climate that is twice as warm as the worst predicted temperature rises. Gaia will regulate the climate outside the city walls. Singapore, he wrote, is not a scorched landscape. It teems with life.[106] The blurb for the book called him "the great scientific visionary of our age."

Lovelock published the book at age ninety-four and in it he described the joys of his fifty-six-year career as a lone scientist. Yet the lone scientist was almost extinct and is not popular in a scientific enterprise organized around massive teams of researchers: what does it say about their work, he asked, when a single scientist can invent the electron capture detector that helped change the course of modern environmental history?[107]

A CELEBRITY MAVERICK

Lovelock showed how celebrity became a weapon for a maverick scientist in the fight to get his ideas heard. Pushed to the edge of science, dismissed as a purveyor of pseudoscience, he spread Gaia through popular culture. The grand idea hit a cultural nerve and came to have deep meaning for a diverse collective of New Age seekers, traditional believers, radical environmentalists, and scientists. As the father of Gaia, he became a star.

With Gaia a key text in the burgeoning ecological movement of the long sixties, Lovelock personified ecology. He manifested in one man the apparently contradictory ideas in ecological thought, between science and antiscience, rationality and spirituality, the material world and the natural world. He perfectly embodied these ideas, the lone scientist with spiritual yearnings working in splendid rural isolation surrounded by acres of woodland he had planted.

Fears over the threat of climate change led to a resurgence of researcher interest in big-picture systemic ways of thinking about the Earth, and Gaia had a newfound relevance. Gaia became climate change personified. Lovelock's ideas—sustained in popular culture through public demand and interest—were welcomed into the scientific mainstream as researchers shifted their scientific work to include his once-derided ideas.

Yet his fame has been a reluctant fame. As *New Scientist* noted in 2013, with Gaia, "Lovelock became a celebrity—though he would probably have traded that for a little more respect from his fellow academics."[108]

EIGHT

The Charming Stardom of Brian Greene

In his New York apartment, physicist Brian Greene removes the two clear protective covers from his first edition of Albert Camus's *The Myth of Sisyphus*, and reads aloud its arresting first sentence: "There is but one truly serious philosophical problem, and that is suicide." Greene continues to quote the French existential philosopher's 1942 essay on the futile search for life's meaning. "All the rest—whether or not the world has three dimensions . . . comes afterwards." Closing the book, Greene, professor of mathematics and physics at Columbia University, said: "I'm no Camus; he's a great, brilliant mind . . . but how can you judge whether life is worth living if you don't know what life—or, more broadly, what reality—is? And to me, the question of whether reality has three dimensions—that is, the dimensions you can see—or that it might have other dimensions . . . that to me adds a richness to life. It makes me want to live."[1]

The physicist opened his home in 2011 to the radio talk show *Science Friday*, as part of the show's Desktop Diaries video feature, where prominent scientists give a guided tour of their workplaces. Dressed in jeans and a white T-shirt under a gray zip-up sweater, Greene was filmed working at a large clutter-free desk on which were two open laptops and a desktop computer. Looking out over the Hudson River, he wrote neat equations in pencil in a large notebook. "There is an idea that genius operates in chaos," he said. "I'm not a genius. I just think better if nothing is around."

Clutter migrated to his bookshelves, he said, where his rare copy of Camus's book sits close to a glass ornament he identified as a Calabi-Yau manifold, a geometric structure that looks like a large knot. "It's the shape of the extra dimensions in string theory," he explained, in reference to his specialist field of theoretical physics, which postulates that the fundamental elements of reality are subatomic strings. The theory argues that the universe contains several hidden dimensions that can be visualized as various structures, such as the one on Greene's shelf, structures that he believes are "actually quite spectacular."

The three-minute, thirty-second clip was a clever strategic presentation of Greene. Nothing random appeared in the short film; even its score was written by Greene's musician father, whom the physicist showed playing the harmonica in a black-and-white video clip on his smartphone. It combined implicit and explicit references to order, beauty, art, science, and the search for a deeper reality—all ideas central to his scientific specialism and his public persona.

At the time it was broadcast, Greene was among the most recognizable public scientists in the United States and abroad. *USA Today* called him "the next Carl Sagan, capable of straddling that chasm between the laboratory and the living room."[2] The *New York Times* labeled him "the cutest thing to happen to cosmology since the neutrino."[3] *Discover* called him "the public face of string theory."[4] The eminent string theorist John Schwarz of Caltech called him "something of a media celebrity."[5]

Greene, cofounder and director of Columbia's Institute for Strings, Cosmology, and Astroparticle Physics, came to public prominence as string theory became the dominant but controversial approach to turn-of-the-century theoretical physics. He also represents a dramatic, unprecedented point in the history of science where at stake is a fundamental reconsideration of what constitutes reality, and what counts as a valid scientific theory about the nature of the physical world. He is an eloquent explainer of the bizarre ideas that have captured the public imagination, as black holes did in the late 1970s. Greene also embodies, more so than any of the other scientists profiled here, the modern telegenic scientist whose public status relies on his ability to glide between the worlds of leading-edge research and popular entertainment without sacrificing his scientific authority. Viewed as a whole, his career shows him to be a new breed of public scientist.

STRING THEORY REVOLUTIONS

String theory is a science of strangeness. It argues that the fundamental unit of reality is not the atom, but is instead a vibrating string, a strand of energy that exists on the scale of 10^{-33} centimeters, known as the Planck scale—"to the size of an atom as an atom is to the size of the solar system."[6] The theory originated in the late 1960s at CERN, the vast high-energy physics laboratory located beneath France and Switzerland. Researchers believed it was a novel technique to explain the puzzling way particles interacted in the nuclei of atoms. From the mid-1970s, some physicists viewed the theory as a way to reconcile the three fundamental forces that govern nature—electromagnetism, the strong force that binds the nucleus of an atom together, and the weak force that causes radioactive decay—with gravity, an elusive reconciliation that frustrated physicists for decades.

But the handful of researchers who worked on string theory made slow progress. They found it riddled with mathematical inconsistencies that meant its calculations did not work out—a major problem that hobbled any new idea in theoretical physics. They found that in order for string theory's equations to make sense, there had to be extra dimensions of space, perhaps seven, in addition to the three dimensions that are already known: length, breadth, and height. More than that, the extra dimensions, the theorists argued, were curled up too tight to observe. By the early 1980s, these problems meant string theory was a research backwater.

That changed in 1984 after string theory's first revolution. American physicist John Schwarz and his British collaborator Michael Green labored on the theory for years and finally published what became a landmark paper that essentially resolved the obstinate mathematical inconsistencies that blocked string theory's progress. The physics community started in earnest to see that string theory, or superstring theory as it was also called, had the potential to solve the greatest problem in modern physics, the problem that Einstein still tried to solve as he lay on his deathbed, the problem whose solution Hawking said would end theoretical physics, the problem of how to integrate gravity into quantum mechanics and therefore merge under a single theory general relativity and quantum mechanics.

Theoretical physicists flocked to the field. The number of papers published on the topic exploded from fewer than a hundred between 1975 and 1983 to twelve hundred in 1987 alone. Physicists glimpsed in string theory what the historian of physics Helge Kragh called "the near-fulfillment of a century-old dream, the stepping stone to a new physics, the sought-after holy grail of a quantum theory of gravity."[7]

Amid this surge of scientific interest, popular writings spread strings through culture. Michael Green wrote about the theory in 1986 for *Scientific American*. Its fundamentals were described by Michio Kaku in *Beyond Einstein* (1987), F. David Peat in *Superstrings and the Search for the Theory of Everything* (1988), and Paul Davies and Julian Brown in *Superstrings: A Theory of Everything?* (1988). The flurry of popular accounts, argued literary scholar Sean Miller, occurred because string theory upended commonsense understandings of the world and because it continued the cultural trend in the twentieth-century Anglo-American cultures where citizens viewed physicists as custodians and explainers of the universe's truths.[8]

As the first string revolution took hold, Greene graduated as a physics major from Harvard in 1984. Afterward, he studied for his doctorate in physics as a Rhodes Scholar at Oxford. It was here that strings first captivated Greene. As he tells it, he passed a bookstore with a poster in the window that advertised a talk by Michael Green, the physicist who helped forge string theory's first revolution. Greene decided to devote his career to it. And he proved to be amazingly good at it, too, joining the physics faculty at Cornell in 1990. Around this time, he made his first signature contribution to the field. He and a colleague showed that every possible shape of the cosmos has a sort of mirror shape that created an alternative universe with the same features as our own.

His second signature discovery came in 1993. With two colleagues, he demonstrated the fabric of space could tear—a rupture prohibited by general relativity. Greene later chronicled the months of calculations that led to the insight that the universe could change shape as it ripped and reformed in a new way. He recalled the moment a sophisticated computer program confirmed that their equations worked out: "I jumped out of my chair and ran an unrestrained victory lap around the office."[9]

But as Greene forged his reputation, prominent physicists lambasted string theorists' claims that their field was actually a science. Their core criticism was that string theory was almost entirely divorced from experiment and observation, the centuries-old ways that scientific ideas were tested, verified, or discarded. But string theory, its critics argued, could not be tested or observed, proven or disproven. No experiments suggested new theories. No theories suggested experiments. "The theory depends for its existence upon magical coincidences, miraculous cancellations and relations among seemingly unrelated (and possibly undiscovered) fields of mathematics," wrote Harvard physicists Paul Ginsparg and Nobel Laureate Sheldon Glashow in 1986. "Superstring sentiments eerily recall 'arguments from design' for the existence of a supreme being. Was it only in jest that a leading string theorist suggested that 'superstrings may prove as successful as God?'"[10]

In 1988, Glashow continued his attack on string theory and theorists. "Should they be paid by universities and be permitted to pervert impressionable students?"[11] he asked, and added: "Are string thoughts more appropriate to departments of mathematics or even to schools of divinity than to physics departments?" Other reasons explained why physicists' initial excitement faded. The mathematics were too tough. The theory's pioneers were too successful at solving major problems.

But another major problem remained: five different, valid string theories existed. Each shared features with the others, but presented different pictures of reality and therefore undermined the field's claims to find fundamental features of the natural world. All the theories cannot explain the essence of reality. Physicists lost faith.[12]

That changed in 1995 with string theory's second revolution. The doyen of the discipline, Edward Witten at Princeton's Institute for Advanced Study, proposed a radical new idea. The five string theories, he argued, were just various ways of describing one master theory: M-theory. String theorists now had an overarching framework for their research. Witten sparked a frenzied interest among theoretical physicists in string theory. It now occupies a central, dominant position within the field, with about a thousand active researchers who together publish thousands of papers. Witten was vague about what M stood for: it has been called magic, mystery, membrane, matrix, or mother.

Yet string theory's controversial scientific status remained. The science writer John Horgan in *The End of Science* (1996) viewed string theory as an exemplar of what he termed "ironic science," research so removed from experimental or observational verification, so open to interpretation, that it was akin to postmodern philosophy or literary criticism.[13] For a time at Harvard a string theory seminar was formally titled "Postmodern Physics."[14] M-theory was also called murky and monstrous.

The conflict over string theory marked an unprecedented moment in the history of science. String theory disturbed traditional techniques of doing physics. Experiments guided and constrained theories. Observations eliminated, supported, or suggested theories. But without the possibility of experimental tests, string theorists turned to mathematics to support their ideas. Now the proof of whether or not a theory explains reality is whether or not its equations work out. The yardsticks of proof are mathematical notions of beauty, elegance, and consistency.

As the historian of science Peter Galison wrote: "What the late twentieth-century is witnessing in the string controversy is a profound and contested shift in the position of theory in physics."[15] And Stephen Hawking and Leonard Mlodinow added in *The Grand Design*: "We seem to be at a critical point in the history of science, in which we must alter our conception of goals and of what makes a physical theory acceptable."[16]

For physicist Lee Smolin, string theory was such a high-risk venture that, if it turns out to be right, "string theorists will turn out to be the greatest heroes in the history of science," but if they are wrong, "then we will count string theorists among science's greatest failures . . . a cautionary tale of how not to do science, how not to let theoretical conjecture get so far beyond the limits of what can rationally be argued that one starts engaging in fantasy."[17]

At the core of this conflict, at this scientific precipice, stands Brian Greene.

A WATERSHED BOOK IN THE PUBLIC UNDERSTANDING OF STRING THEORY

Greene pushed this clash of ideas into public view. *The Elegant Universe*, first published in 1999, described how the universe worked according to string theory. He positioned the field as a seamless continuation of Einstein's unfulfilled search to uncover a unified theory that would show how all phenomena of physics are just "reflections of one grand physical principle, one master equation." Greene wrote: "From one principle—that everything at its most microscopic level consists of combinations of vibrating strands—string theory provides a single explanatory framework capable of encompassing all forces and all matter."[18]

The idea captured end-of-the-millennium public imagination and the book sold hundreds of thousands of copies, outranked for three days on Amazon the just-published John Grisham thriller, was a finalist for a Pulitzer Prize, was named a notable *New York Times* book, and won the British Aventis Science Book Prize. One scholar called it a "watershed moment" in the popularization of string theory.[19]

Greene had several rhetorical aims with *The Elegant Universe*. First, he offered nonexpert readers an insider's view of the latest trends and thinking in theoretical physics, a field with an established popular science audience for intelligent treatments. The book, Greene explained in the introduction, grew out of general-level lectures he had given on relativity, quantum mechanics, and string theory. In these lectures, Greene said, he sensed the audience's enthusiasm for the ideas, even though the first literary agent he approached rejected the proposal outright, believing it was unlikely to reach a mainstream audience. Second, Greene aimed the book at science students and teachers and he included extensive technical notes for reference. And third, he also wrote to present to scientists in other fields a fair and honest account of why string theorists are so excited.

Critics praised Greene's clear prose. Journalist George Johnson, author of *Strange Beauty: Murray Gell-Mann and the Revolution in Twentieth-Century Physics*, wrote in the *New York Times*, for example, that describing mathematics for readers assumed to have no background in the subject is the hardest challenge in popular science writing, but Greene did it "with a depth and clarity I wouldn't have thought possible." He wrote that: "Greene lays out the theory so clearly and persuasively that you almost come to believe it, and to hope that someday a way will be found to establish whether it is anything more than wishful mathematics."[20]

But others found the writing too dense with technical details. Eminent science writer Marcia Bartusiak, reviewing the book for the *Washington Post*, noted the book's strength was its translation of mathematics into visual terms. "Yet his desire to reach the general reader may be overly ambitious. His discussions of gauge symmetries and Calabi-Yau geometries will be best appreciated by the science-minded who seek an insider's perspective on the cutting edge of physics."[21] The science editor of the *Observer*, Robin McKie, found the book "long, dense and extremely technical."[22]

Criticism and promotion tied Greene's writing style to his science. The back-cover blurb of the 2003 Vintage edition of the book, for example, said the "writing [is] as elegant as the theories it explains." Reviewing the book for the history of science journal *Isis*, physicist Laurie M. Brown argued that Greene's "writing is clear, entertaining, sometimes humorous or poetic—in a word, elegant."[23] Analyzing his book for the journal *Public Understanding of Science*, literary scholar Rachel Edford wrote: "Beauty, harmony, and elegance are important thematic as well as stylistic elements in Greene's text."[24]

Beauty is central to string theory's scientific status. Physicists have particular definitions of beauty and elegance—and these terms help assess whether a scientific theory is true or not. As Nobel physicist Steven Weinberg explained, an elegant theory contains no irrelevant detail but has enormous power to explain things. A beautiful theory is simple and its reasoning is so clear and logical that its conclusions seem inevitable.[25] In *The Elegant Universe*, Greene wrote that "some decisions made by theoretical physicists are founded upon an aesthetic sense—a sense of which theories have an elegance and beauty of structure on par with what we experience." But he noted: "Ultimately, theories are judged by how they fare when faced with cold, hard, experimental facts."[26]

Expert commentators showed that Greene's popular writing had a scientific agenda. Greene is "a proselytizer of the cause"[27] of string theory, wrote former *Nature* editor John Maddox. "Greene is a self-confessed cheerleader for superstring theory," wrote science writer Bartusiak, "which can leave the wrong impression that this is the only route to a theory of everything." Rachel Edford noted that Greene did not mention string theory's main competitor, loop quantum gravity, which consists of a different approach to unification.[28] And John Schwarz—who sparked the first revolution in the field—noted in his *American Journal of Physics* review that Greene's account was "very much a personal view of a large, complex, and rapidly evolving field."[29]

Nevertheless, Schwarz and Maddox viewed Greene and the book as catalytic for work in the field and for public support for its work. For Maddox, much work needed to be done in the field, he said, and the "thousand and more people working in the field will need courage to do these largely thankless chores, but this splendid book will cheer them on their way." For Schwarz, Greene's "youthful vigor (and acting experience) serve him well. . . . In my opinion, the publicity that he gives the field is a very healthy development." Greene, he wrote, had become "something of a media celebrity."[30]

Greene's celebrification grew out of the publicity and promotion around *The Elegant Universe*. A week before its publication, *New York* magazine, the chronicler of cultural trends, published a profile of Greene. The magazine called him "boyish" and "mediagenic." It said his publisher, W. W. Norton, "hopes he can do for string theory what Stephen Jay Gould did for evolution, Stephen Hawking for black holes, and Richard Feynman for quantum electrodynamics: give science a friendly face, and make a chart-busting best-seller out of a rarefied subject."

The article painted Greene as an anomaly in professional physics: "a striking standout, and it's not just because he's attractive and wears contacts. In high school, Greene won math competitions *and* judo tournaments. At Harvard, he performed in musicals. At Oxford, he hung out with George Stephanopoulos," the political analyst and former aide to President Bill Clinton. It cited a Columbia math graduate student: "He's a great communicator, he's charismatic, he's clearly top-of-the-heap intellectually. So the fact that he has gobs of raw physical appeal on top of that—it gives him a really serious mystique."[31]

The profile enhanced that mystique with personal details. His father was a composer, voice coach, and vaudevillian. As a boy he was a math prodigy who multiplied thirty-digit numbers by thirty-digit numbers. As a grade school student, he was sent by a teacher to Columbia's Mathematics Department with a note asking for someone to help him develop his talents. As an adult, his girlfriend at the time, Ellen Archer—"she's an actress, naturally"—recounted how Greene, while driving, became so lost in thought that the car would slow almost to a stop.

But journalists also picked apart Greene's constructed media image. "Greene promotes himself as a happy-go-lucky kind of polymath,"[32] wrote the *Australian*. "He is an unusually telegenic physicist." The *Christian Science Monitor* called him a "dashing math and physics professor . . . who is luring lay people into learning about cutting-edge physics with his engaging prose and soothing, late-night-radio-host voice."[33] *USA Today* wrote that, despite string theory's complexity, Greene "has just the right mix of scientific brilliance and telegenic appeal to sell it as physics chic."[34]

Other publications profiled Greene as the hottest researcher in the hottest field in physics. *Scientific American* said his explanatory prowess and "youthful good looks" meant he quickly became "the poster boy for theoretical physics." The magazine quoted Stephanopoulos, who joked that "Greene, who is single, might be the first physicist to have groupies." It said that, after Boston and Oxford, "Manhattan seems the perfect place for Greene. Partial to chic black clothes, he bears a slight resemblance to the actor David Schwimmer of the TV series 'Friends,' with the same boyish charm and comic timing. Only a touch of gray in his wavy hair hints that he is 38." Since *The Elegant Universe*, the magazine wrote, "Greene has gotten the buzz as [string theory's] hottest practitioner, his fame eclipsing even that of Edward Witten."[35]

Newsweek described him as "articulate, witty and totally nongeeky (black jeans, contacts, former wrestler, vegan)" and "[l]ow-key, thoughtful and adamant about not being seen as the face of strings."[36] "I think people are struck by the ideas," Greene told the magazine. "I don't think they are struck by me."

SEEING THE STRANGENESS OF STRINGS

Greene used his fame to spread string theory's ideas through a powerful medium—film. He supplied technical dialogue for the physics protagonist of the offbeat 1990s television comedy *3rd Rock from the Sun*, but he moved in front of the camera in 2000 when he appeared in a cameo role in the film *Frequency*. The movie relied on the premise that a police officer in 1999 can talk via a radio to his dead firefighter father when he was alive in 1969, a dramatic foundation built on ideas of alternative histories and time travel.

Greene appears as "Prof. Brian Greene," a physicist during the pivotal scene where father and son first speak to each other over a ham radio. As an aged version of himself on a color television in 1999 and a younger version on a black-and-white television in 1969, he discusses how string theory might lead to a dramatic new understanding of space and time. He explains how string theory requires our universe to have ten or eleven dimensions and that physicists are also examining the idea of multiple alternative universes.

The physicist also worked as a technical consultant on *Frequency* and several film reviewers commented on the scientific ideas. The late eminent critic Roger Ebert wrote: "Of course the latest theories of quantum physics speculate that time may be a malleable dimension, and that countless new universes are splitting off from countless old ones all the time."[37]

He was one of a new breed of science advisors to the entertainment industry. These consultants advised the filmmakers how to make the movie's science accurate and make its ideas look real. But, as science studies scholar David Kirby argued in *Lab Coats in Hollywood* (2011), science consultants like Greene promote their science through movies. They help present controversial ideas in dramatic stories, and this presentation suggests that these concepts, instead of being contested scientific ideas, represent reality. By presenting the ideas in popular culture, science consultants position their ideas as worthy of increased public and scientific attention. Movie storylines affect scientific work.

But the primary reason producers hired Greene as a consultant, argued Kirby, was that the best selling author's fame helped promote the film. "Filmmakers, however, are not hiring a scientist like Brian Greene to have him devote weeks fixing scientific details," wrote Kirby. "They are primarily hiring Greene because [of] his bestselling 1999 book . . . and he has been favorably compared to Carl Sagan because of his good looks and easy manner on television." Greene exemplifies the fact that science consultants' "primary value rests on their scientific expertise transferring to a film through their 'celebrity endorsement.'"[38]

Greene also brought his ideas to television. He repackaged *The Elegant Universe* for a three-part *Nova* series, also called "The Elegant Universe," which took two years and $3.5 million to produce. The show's executive producer, Paula Apsell, described Greene as a "sexy, smart scientist."[39] Television lent itself to dramatic presentation of quantum mechanics and string theory. It was a perfect topic for *Nova*, too, as it allowed the construction of dramatic visual images to attract its core audience of elite viewers interested in science. As Dennis Overbye wrote in the *New York Times*, the medium permitted the visualization of weirdness. The show was a spectacle of virtual reality. The viewer sees Greene play the cello to explain string vibrations, slice bread to show parallel universes, leap from a tall building and land on his feet to show electromagnetism's effect on gravity.

"String theory says we may be living in a universe where reality meets science fiction," intoned Greene as host. The dramatic visualization of string theory made it impossible to separate material reality from the razzle-dazzle of special effects. As one scholar noted, "The Elegant Universe" "presents a multimedia spectacle which magically turns . . . speculation into proven claim."[40]

Reviewers disagreed about the merits of the show's sci-fi style—but agreed Greene was an engaging host. Dan Vergano in *USA Today* said it succeeded in showing how physicists viewed reality. He noted that "Greene is an arresting speaker, introduced at one physics conference last year as a 'rock star of physics.' But he plays it low-key."[41] The *Sunday Times* focused on Greene as host. "He is conventionally attractive (tall, dark-haired, dresses in the comforting casual-Friday-style of modern academia)," it wrote, and is "enthusiastically determined to explain his subject to as many people as possible."[42]

However, Virginia Heffernan in the *New York Times* criticized the "daffy presentation of the show—big glops of animation and surrealism accompanied by blurpy sound effects." *Nova*, she argued, wanted "to show viewers that string theory is cool" and to do this, it relied "on the force of Mr. Greene's personality even more than its glossily produced interludes of surrealism." He is persuasive, she wrote, because as "he seems to pirouette through the odd hallucinations, talking like a maniac about how dang amazing string theory is, it becomes increasingly impossible to doubt that he's feeling it." Greene, she concluded, is "a compact and antic man who has participated in musicals and judo tournaments, [who] emerges as a motivational guru, a Tony Robbins of physics."[43]

THE EASY SELLING OF PROF. GREENE

By the time the shows were broadcast, Greene's commercial value had soared. At the 2000 Frankfurt Book Fair—the major annual industry event where publishers compete to buy new works—Knopf paid $2 million for Greene's unfinished next book. (The show business memoirs of former pop star Victoria Beckham, by contrast, fetched 1 million pounds.) The *Times* argued that two factors drove up the price: "a justifiable belief that Greene was the new Hawking, only better, and the fact that he looked good—or, at least, not like a scientist."[44] As a spokesman for Random House (Knopf's parent company) told the paper: "Professor Greene is brilliant and wonderfully lucid, and tremendously photogenic."[45]

That unfinished book became *The Fabric of the Cosmos* (2004), which explored the question: "What *is* reality?" Greene argued that "just beneath the surface of the everyday *is* a world we'd hardly recognize" and showed readers this veiled world through an examination of how the scientific understanding of space and time—the fabric of the cosmos—evolved over centuries through the work of Newton, Einstein, Bohr, and Heisenberg. "If superstring theory is proven correct," he wrote, "we will be forced to accept that the reality we have known is but a delicate chiffon draped over a thick and richly textured cosmic fabric."[46]

Reviewers responded to Greene's obvious passion for physics. Not atypical of the critical response was the *Jerusalem Post*'s characterization of the book as "a model of enthused exposition."[47] That enthusiasm overcame, for Janet Maslin in the *New York Times*, the high-level technical content. She wrote: "his excitement for science on the threshold of vital breakthroughs is supremely contagious."[48]

Commentators analyzed Greene's persona as much as his prose. The *Washington Post* noted: "His reputation as a leader in the field of 'string theory' must inevitably compete with his reputation as a good-looking guy who's comfortable on television."[49] The *Herald* called him "a TV producer's dream of a scientist. Youngish and photogenic, Greene looks likes he has just stepped out of The X-Files."[50] The *Irish Times* opened its interview with: "When he laughs he looks like John Cusack; when he's serious there's a hint of David Duchovny."[51] The *Globe and Mail* noted that "as befitting someone who obsesses on the nature of a cosmos that is 90 percent composed of dark matter and dark energy," he "is dressed entirely in black, save for the peek-a-boo flash of a white T-shirt." Greene reflected (in another interview) on those comparisons with handsome actors. He joked with one interviewer that he must have a "generic face."[52]

The Fabric of the Cosmos reignited the dispute over string theory's scientific status. For eminent physicist Freeman Dyson, string theory was an example of a debate that recurred in the history of science—a clash between revolutionaries and conservatives. Classifying himself as a conservative, "out of touch with the new ideas and surrounded by young string theorists whose conversation I do not pretend to understand," Dyson said that he doubted string theory's value as science—it was one of several "fashionable theories"—because it was not currently testable.[53] Another physicist said Greene should not publicly discuss ideas that are not yet proven. When Greene discusses string theory, wrote the physicist, he "too often makes judgments based on what he calls his 'gut' feelings. Theories that have not advanced beyond such a primitive first stage in the minds of scientists should not be paraded outside a circle of experts."[54]

A challenge to string theory in popular science followed in 2006. Two scientists wrote books that said string theory damaged physics. Lee Smolin, the developer of loop quantum gravity, the rival unified theory to string theory, argued in *The Trouble with Physics* that the theory has not fulfilled its early promise, yet it continued to monopolize talent and resources, draining life from other subfields. String theories labor on a flawed idea, he argued, because a form of groupthink grips the field; a tight-knit collective of competitive scientists believe it is impossible they are wrong. Physicist-turned-mathematician Peter Woit argued in *Not Even Wrong* that not only was string theory an unlikely candidate for a theory of everything, it did not even deserve the status of a candidate theory. It was so wide of the mark that—in the words eminent physicist Wolfgang Pauli used to describe a useless scientific theory—it was not even wrong.[55]

But the field, Woit wrote, has been "spectacularly successful on one front—public relations." The elements of this strategic communications campaign, he argued, were *The Elegant Universe*, NSF-supported *Nova* shows about the book, and regular *New York Times* articles and conferences to train teachers to educate students about strings.

Both scientific critics argued that, to redress the overemphasis on string theory, the sociology of physics needed to change. Senior physicists should switch their research to less popular areas, and encourage their students to do the same. "I would argue that a good first step," Woit wrote, "would be for string theorists to acknowledge publicly the problems and cease their tireless efforts to sell this questionable theory to secondary school teachers, science reporters and [grant] officers."[56]

(Incidentally and not insignificantly, Smolin wrote elsewhere: "I get mail from readers who complain that I am not as good-looking as Greene, even though I write better."[57])

Reviewing both books for *American Scientist*, string theorist Joseph Polchinski said "scientific judgment" is the main influence that gives string theory its glamour, promise, and relevance.[58] If there were nothing significant to be found in string theory, he argued, these accomplished scientists would migrate to more promising areas. A value of these books for one philosopher of science is that, through them, "the conflict about the status of string physics is clearly visible to a wider public."[59]

SPREADING SCIENCE THROUGH CULTURE

Meanwhile, Greene continued to communicate science—in all directions.

In 2011 he hosted "The Fabric of the Cosmos," a four-part *Nova* show based on the book. The show was another spectacular multimedia spectacle that once again made visual, in a sense made real, ideas from the edge of physics. In a still from the show, we see a glittering vision of the universe appear before Greene as he sits late at night in "Albert's Diner," and explains Einstein's contribution to our knowledge of a mysterious phenomenon that shrouds the universe: dark energy.

Greene worked as a science consultant on the film *Déjà Vu* (2006), produced by blockbuster creator Jerry Bruckheimer and directed by the late Tony Scott. In the film, actor Denzel Washington plays a government agent who uses secret technology to travel back in time through a wormhole to prevent a terrorist attack that slaughtered civilians. The film explores parallel universes and the paradoxes of time travel, and features a string theorist character.

Greene has also allowed himself to be lampooned. On the hit TV sitcom *The Big Bang Theory*, he gives a bookstore talk. There he tries to explain Heisenberg's uncertainty principle, a foundational idea in quantum physics,

Once called "an unusually telegenic physicist," Brian Greene used special effects to visualize the texture of the universe in PBS's "The Fabric of the Cosmos," a 2011 show based on his best selling book of the same title. AP Photo/PBS/NOVA and Pixeldust Studios

where either the speed or the position of a particle can be measured, but not both, saying it's "much like the special order menu that you find in certain Chinese restaurants where you have dishes in column A and other dishes in column B, and if you order the first dish in column A, you can't order the corresponding dish in column B." Main character Dr. Sheldon Cooper bursts out laughing and asks Greene: "You've dedicated your life's work to educating the general populace about complex scientific ideas. . . . Have you ever considered trying to do something useful? Perhaps reading to the elderly? . . . But not your books—something they might enjoy?"[60]

He also appeared on the late-night comedy show *The Colbert Report*, which ribbed him and his work in an interview. Stephen Colbert summarized string theory as: "It's multidimensional, vibrating strings that are so small that there's no way to ever measure whether they exist, or that the state of the universe that you describe in your mathematics can be quantified in physical, verifiable ways on any level, at any imaginable point in the future and yet . . . people still take you seriously."[61]

Greene also brought physics to younger audiences. His children's book, *Icarus at the Edge of Time* (2008), tells the story of a boy who builds a spaceship and flies close to the event horizon at the edge of a black hole, where conventional notions of time shatter. When he returns from his journey, it is 10,000 years later. The book became the basis for an artistic collaboration, as he transformed the book into an orchestral work in collaboration with the composer Philip Glass, who previously wrote operas about Kepler and Einstein. Greene discussed his family life when promoting the book. "At first I didn't want to force the book on my five year old. Only recently did my wife read it to him, and he cried at the end."[62]

The *New York Times* published several op-ed articles Greene wrote, one of which articulated most clearly his philosophy of science communication. He told how a soldier in Iraq had written him a letter and described clinging to one of his books amid the chaos, because it showed how all humans are part of a grander cosmos. For Greene, the letter spoke "to the powerful role science can play in giving life context and meaning."

"At the same time," Greene continued, "the soldier's letter emphasized something I've increasingly come to believe: our educational system fails to teach science in a way that allows students to integrate it into their lives." It was necessary, he argued, to "cultivate a general public that can engage with scientific issues; there's simply no other way that as a society we will be prepared to make informed decisions on a range of issues that will shape the future." To do this, he argued: "We must embark on a cultural shift that places science in its rightful place alongside music, art and literature as an indispensable part of what makes life worth living."[63]

He put this philosophy into practice. Inspired by an event he attended in Genoa, Italy, he and his television producer wife, Tracy Day, in 2008 created the World Science Festival in New York. More than one hundred twenty thousand people attended the five-day festival to see one hundred thirty performers and speakers, among them eleven Nobel laureates. Greene and Day hope "to sustain a general public informed by the content of science."[64] Greene did not want to present a childish, dumbed-down science. "Some people think the only way the general public will engage with science is to make it 'oh, so fun,'" he said, "with balloons and explosions and confetti. I really hate that. The public will engage with the ideas of science—you just can't make it boring."[65]

In *The Hidden Reality* (2011) Greene explained the multiverse, the idea that our universe might be just one of billions that exist, each with its own particular characteristics. He conveyed the varieties of the multiverse (or metaverse or megaverse). There are parallel universes where copies of everything in this universe exist simultaneously in another reality, some hidden millimeters from us while others are removed by massive stretches of space and time. The braneworld scenario posits that our universe is but one of several sheets, or branes, that float in other dimensions of space. These branes, in another variation, slam into each other and then push apart over and over again, each time creating new universes in an eternal cycle. The landscape universe holds that all the shapes that contain the extra dimensions in string theory exist together on a single, massive terrain. The holographic universe argues that our world is nothing but a projection of a reality that takes place elsewhere. The ultimate multiverse, the final idea described in the book, claims that every potential universe exists somewhere in a grander, all-encompassing multiverse.

For Greene, the ultimate multiverse is the most fanciful version.[66]

The publication of *The Hidden Reality*, along with the broadcast of "The Fabric of the Cosmos," led to the most journalistic attention Greene has yet received (see chart).[67] The multiverse captured public attention as the scientific-idea-of-the-moment. A string of popular science titles, including Michio Kaku's *Parallel Worlds* and Hawking and Mlodinow's *The Grand Design*, explored the exotic idea. And parallel universes and alternative histories and other dimensions gained traction, Greene noted, because general readers were already exposed to fictional representations of these concepts. He cited as his favorite personal examples the films *The Wizard of Oz*, *It's a Wonderful Life*, *Sliding Doors*, and *Run Lola Run*, as well as the *Star Trek* episode "The City on the Edge of Forever" and the Jorge Luis Borges story "The Garden of Forking Paths." Greene wrote about the multiverse in a *Newsweek* cover story and also spoke about it at a TED talk.

Critics found *The Hidden Reality*, with its thirty pages of technical footnotes, a dense read. Janet Maslin, a longtime Greene observer, wrote in her *New York Times* review: "It's exciting and rewarding to read him even when the process is a struggle. This book is significantly more difficult than his earlier ones, but it still captures and engages the imagination."[68] Physicist George Ellis in *Nature* said Greene slid from science to philosophy, as he did not present facets of reality, but instead told of "unproven theoretical possibilities."[69]

Countering this type of critique in a *New Scientist* interview, Greene said: "I think it's important for the general public not just to learn about science that's all settled, confirmed and in textbooks, but also to capture a picture of vital science in the making. That's the stage we're at now when it comes to the idea that our universe may be one of many."[70]

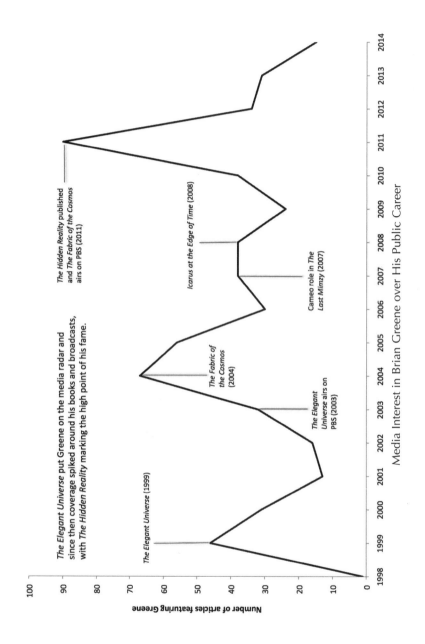

Media Interest in Brian Greene over His Public Career

The Elegant Universe put Greene on the media radar and since then coverage spiked around his books and broadcasts, with The Hidden Reality marking the high point of his fame.

The Hidden Reality published and The Fabric of the Cosmos airs on PBS (2011)

The Elegant Universe (1999)

The Fabric of the Cosmos (2004)

The Elegant Universe airs on PBS (2003)

Icarus at the Edge of Time (2008)

Cameo role in The Last Mimzy (2007)

Number of articles featuring Greene

The ideas in Greene's books merged with spiritual themes. Rabbi Yitzhak Ginsburgh, for example, argued the multiple dimensions of reality proposed by string theory evoked similar concepts in Kabbalah, a form of Jewish mysticism.[71] Greene—who was born Jewish but who is not religious—is aware of the spiritual and metaphysical interpretations of strings. His half-brother, Joshua, is a Hare Krishna devotee, and Greene often told him about new ideas in physics. "And a lot of times, he'll say to me, 'Well, we already knew that. That's in Vedic text number 23 or something of that sort.' And, you know," said Greene, "it's an interesting exchange because when we've ever gone into more detail on it, it seems as though many of the ideas that we have come upon do have a kind of resonance with ideas that have been articulated in ancient text or even in more modern theological or mystical text."[72] Elsewhere Greene remarked: "So, from my dad came two seekers. One went the religious route, one the scientific route."[73]

But Greene deftly sidesteps the inevitable science-and-religion debate. In interviews, he has said he does not see science and religion in conflict, except in the narrow context of teaching creationism instead of evolution. "My own personal view is science can't rule out religion. It can't rule out God. God could be behind it all, set it all up so we've discovered what we've discovered but set it up so that we don't have any direct evidence for his existence or her existence, whatever the right word is," he told the *Daily Beast*. "My view is if what we are doing as scientists is working out God's design, and that's what we're doing, if that's the case, I'm thrilled to be part of that momentous journey. But if that's not the case—and I don't think it is—if what we're doing is just working out the deep laws of the cosmos that brought the universe into existence and have guided its evolution, if I'm part of that journey and I can take that a little bit further, I'm thrilled to be part of that. So, in a day-to-day way, it just doesn't matter."[74]

Yet these strange, speculative ideas are integral to Greene's popular appeal. Amanda Schaffer in *Slate* wrote an astute analysis of one dimension of Greene's cultural status, framing her argument around the question of why Greene was a celebrity, but Freeman Dyson was not. The "unselfconscious romanticism" of Greene's writing, she wrote, merged with his presentation of string theory, with the result that the potentially never-falsifiable theory is "a perfect canvas on which idealism, optimism, and romanticism can be easily projected." As a result, Greene was "treated as a kind of New Age, scientific guru by the public."

The piece concluded: "It's easy to see, then, why those inclined toward New Age thinking, or who search for spiritual significance in the material world, would find Greene highly attractive; Deepak Chopra would love Brian Greene (though the reverse is probably not true)," wrote Schaffer. "The very qualities that make fellow scientists skeptical—the obsession with elegance, the quasi-spiritual shtick—are precisely what dazzle a public hungry for meaning."[75]

STRING THEORY STAR

Greene's fame has had a clear trajectory. As string theory flourished in the 1990s, he established his scientific authority at the start of the decade with two major contributions to the discipline, one of which proved one part of Einstein's relativity wrong. He broke into general culture at the end of the century with *The Elegant Universe*, which demonstrated he was a passionate, eloquent explainer of an esoteric subject. The promotion and publicity for the book merged his personality with his work to present a media-friendly image of him as a brilliant, but atypical, physicist. He was a credentialed expert, wrote clearly and persuasively, and had a natural talent for communication in books and on television. Greene became the public face of string theory, its most eloquent, earnest, and excitable salesman.

Greene exemplifies deeper currents in physics and science. He is the most visible advocate for the latest candidate for a theory of everything in physics. He also illuminates debates about what counts as a valid theory about the natural world, one that relies on abstract notions of aesthetic elegance and mathematical consistency over one that traditionally bows to observation and experiment. He illustrates, furthermore, a turning point in the history of science, where theorists have broadened the definition of reality, a definition that goes beyond what can be seen and measured to include hidden dimensions, parallel universes, and alternative histories. His communications reinforce the academic legitimacy of string theory. Greene's work aids the attraction of top students and funders to string theory, draining talent and money from other subdisciplines that are less glamorous, promoted, and visible.

But perhaps the major reason why Greene is such an important figure is the way he pivots between scientific and popular cultures—without losing status in either realm. He is at once an influential scientist, cerebral author, cultural impresario, and popular television host. He brings students to the field and brings the field to society. His workplace is not just the classroom and lecture hall but also the venues where today's public intellectual must perform: the newspaper interview, the op-ed, the TED talk, the satirical news show, the late-night talk show. He demonstrates the virtual obsolescence of the Sagan Effect. In its place is a new phenomenon, one where the scientist communicates in nonfiction books, television spectacles, sci-fi, comedy shows, science festivals, and movie cameos—all media appearances that promote a scientist to different audiences. But this communication not only spreads scientific ideas through culture. It uses culture to spread ideas into science. Call it the Greene Effect.

NINE

Neil deGrasse Tyson's Star Quality

Dreams of space animated 1960s America. The Russians' Sputnik 1 became the first satellite to orbit Earth in 1957, sparking a dramatic escalation of the Cold War. The United States created NASA. In 1961, President John F. Kennedy told Congress he wished to send a man to the moon before the end of the decade. That, he envisioned, would be the cornerstone of the national endeavor to "move forward, with the full speed of freedom, in the exciting adventure of space."[1] The photography-driven mass-market *Life* and *Look* magazines crafted the imagery that accompanied this dominant utopian vision, presenting white astronaut heroes as suburban dads and their white spouses as perfect housewives.[2]

Astrophysicist Dr. Neil deGrasse Tyson was born the week NASA was founded in 1958 and a year and a day after Sputnik was launched.[3] But he did not share the excitement of space exploration. "The people NASA was lining up to send into space, in those early years, were many skin shades lighter than I was," he told the *New York Times* in 2007. "It meant that NASA did not have me in mind in this new frontier of space." In 1964, when he and the space agency were both six years old, he saw picketers outside an apartment block in the Riverdale neighborhood of the Bronx, where his family wanted to live.[4] As a result, Tyson wrote: "There was a limit to how much I could celebrate America putting white military pilots into space,

179

while my family was being denied the option to move into the apartment building of our choice."[5]

The black press in the 1960s positioned the space race within civil rights struggles. *Ebony*, the magazine aimed primarily at the black middle class, and other black media viewed the endeavor through the prism of social problems—inner-city strife, unfair housing, inadequate public transportation, and racism. Black journalists contrasted the $4 billion annual expenditure on the Apollo program with the lack of initiatives to combat urban poverty. When Apollo 11 landed on the moon in 1969, an editorial in *Ebony* linked spaceships and slave ships.[6] Tyson wrestled with this history as he described his choice of career: "I am the only one in my generation who became an astrophysicist *in spite* of your achievements in space rather than *because of* them."[7]

Tyson drove himself to become one of the most prominent scientists in the contemporary United States. He is director of the Hayden Planetarium in New York, has written ten popular books so far on cosmology, hosted four seasons of a *Nova* spin-off show, a science radio show, and an updated version of Carl Sagan's *Cosmos*. Tyson has sat on two presidential scientific committees, sits on the advisory board of NASA, and was awarded the space agency's Distinguished Public Service medal, the highest award given by the space agency to a nongovernmental worker who has advanced the work of the agency.

Tyson's influence has reverberated through culture. *Discover* in 2008 named him as one of its ten most influential people in science.[8] The *Washington Post* noted that Tyson has "inherited the job created by Carl Sagan: pop culture's public brain for all cosmos-related matters."[9] *Playboy* said Tyson had "turned himself into a rock-star scientist."[10] *Ebony* noted that Tyson "shines like a supernova in the constellation of scientific celebrities."[11] Tyson became a public figure amid the aftershocks of the 1960s, and went on to significantly influence the public understanding of science, science policy, and scientific debates—even though his record as a researcher remains modest. More so than any other scientist profiled here, he demonstrates the considerable power wielded by a public scientist who is a scientific celebrity.

THE DILEMMA OF THE
BLACK SCIENTIFIC INTELLECTUAL

The civil rights struggles continued into Tyson's college years. He studied physics as an undergraduate at Harvard in the late 1970s. The decade witnessed a breakdown in institutional barriers that had prevented African Americans from attending university. As a result, scholarship by black academics flourished. Programs of black studies, an interdisciplinary approach to knowledge about black people in American history, society, and culture, opened in universities across the country. Tyson graduated in 1980, at a time when black intellectuals such as bell hooks and Henry Louis Gates began to establish themselves in U.S. universities and to achieve a notable, but marginal, presence in American public life.[12]

But black intellectuals in the nineteenth and twentieth centuries faced a particular intellectual tension. These thinkers struggled with their desire to follow their chosen intellectual pursuits while at the same time fulfilling their felt duties to their community and its race-related struggles,[13] a tension articulated by the African American public intellectual Cornel West in an influential 1985 essay, "The Dilemma of the Black Intellectual."[14]

Tyson felt this dilemma acutely. At Harvard, he studied the cosmos, while many of his talented African American peers combined activism and academia in their examination of civil rights–related topics such as housing inequalities, job discrimination, and unfair education.[15] When he was a sophomore, he told another African American member of the college wrestling team, an economics major who planned to assist impoverished communities, that he wanted to be an astrophysicist. His friend replied: "Blacks in America do not have the luxury of your intellectual talents being spent on astrophysics."[16]

In response, Tyson said: "I knew in my mind that I was doing the right thing with my life . . . but I knew in my heart that he was right. And until I could resolve this inner conflict, I would forever carry a level of suppressed guilt for pursuing my esoteric interests in the universe."[17] After he graduated from Harvard, Tyson undertook postgraduate work in astronomy at the University of Texas at Austin, still carrying the feeling that following his scientific vocation meant he "became further isolated from the brilliant good-deed doers of my generation."[18]

At Austin, he specialized in star formation models for dwarf galaxies, small galaxies composed of several billion stars. During his graduate studies, he developed his talent for communication. In 1983, he took over the production of an astronomy column for *StarDate*, a general audience publication of the McDonald Observatory of the University of Texas at Austin. Tyson wrote pseudonymously as the fictional character Merlin. His columns, some of which were published in Tyson's first book, *Merlin's Tour of the Universe* (1989), were explanatory and educational, like telling readers what would happen if the world stopped turning.

In a sequel, *Just Visiting This Planet* (1998), he explained his philosophy of communication: "As a literary vehicle, I have rebuilt the famed Merlin character of Arthurian legend into an educational tool that allows me to explore creative ways of bringing complex topics of the universe within reach of the lay reader," he wrote. "The Merlin character simultaneously embodies my enthusiasm for astrophysics and my daily desire to share cosmic discovery with the public."[19]

After he graduated from Texas with an MA in astronomy in 1983, he began doctoral work in astrophysics at Columbia University in New York, where he specialized in the study of a phenomenon known as galactic bulge, a tightly packed group of stars within a larger star formation. He was also the department's media contact, an informal role that provided him with the means and opportunity to achieve a dramatic and public resolution to the dilemma that remained with him since his undergraduate years.

SHATTERING RACIAL STEREOTYPES

In the late 1980s, a local Fox affiliate called Tyson at Columbia to ask if he would discuss on air the potential consequences that explosions on the Sun's surface would have for Earth. As Tyson would later recollect in his autobiography, it was a call that "would change my life."[20] Watching the segment on television later that night, Tyson said he experienced "an intellectual out-of-body experience." Tyson wrote:

> On the screen before me was a scientific expert on the Sun whose knowledge was sought by the evening news. The expert on television happened to be black. . . . At that moment, the entire fifty-year history of television programming flew past my view. At no place along that timeline could I recall a black person (who is neither an entertainer nor an athlete) being interviewed as an expert on something that had nothing whatever to do with being black.[21]

His appearance as an expert on a non-black-related issue resolved his dilemma of being a black intellectual in science. "One of the major barriers to successful relations between blacks and whites is the latent supposition that blacks as a group are just not as smart as whites," he recounted. "I had finally reconciled my decade of inner conflict. It's not that the plight of the Black community cannot afford having me study astrophysics. It's that the plight of the Black community cannot afford it if I don't."[22]

Tyson's expertise allowed him to escape from the roles that predominantly white gatekeepers in mainstream media organizations typically assigned to black intellectuals. Producers and editors usually expected black thinkers to address black topics. And black experts faced the related problematic expectation that they must present the black perspective or the black voice on an issue, a pattern underwritten by an assumption that all black experience has a single essential nature.[23] But Tyson's appearance on television meant he became entwined in another problematic race-related issue: the media portrayal of African Americans. Black men in particular were often associated with violence and hypersexuality, labeled as violent or criminal or incompetent or uneducated. Black success was portrayed as acceptable only within entertainment and sports. Omitted from these stereotypical representations was the vast experience of working, married, middle-class blacks.[24]

Tyson discussed his life and career against these prevailing stereotypes. "Yes I was good at basketball, and yes, I could slam dunk in ninth grade. I was captain of the wrestling team," he wrote. "I was sort of your quintessential Black athlete. As expected. It was easy. And why was it easy? I'm convinced it was because all the forces in society *allowed* it to happen." But the reaction to his academic accomplishments was different. When he was appointed as editor-in-chief of the physical science journal at his Bronx High School of Science, he recalled: "people started murmuring, 'Well, how did *he* get to be . . . ?' 'What did *he* do . . . ?' All this sort of undercurrent. But no one questioned my athletic achievements. No one questioned that at all."[25] Journalists, he noted, often included coded references to his background. They erroneously placed his roots in the violent South Bronx, an area associated in the popular imagination as "a spectacular set of ruins, a mythical wasteland, an infectious disease."[26] Yet at the same time, Tyson made frequent reference in his autobiography to his physicality and his sporting prowess. He discussed his skills as a college wrestler and once noted that, to earn money in college, he considered working as a male stripper.[27]

Although he was presented in public as an expert in astronomy who happened to be black, racial stereotypes affected Tyson's reception within astrophysics. "When I first entered graduate school (in an institution far from Columbia), I was eager to pursue my dreams of research astrophysics," he noted in the convocation address he gave when he graduated with a PhD from Columbia in 1991. "But the first comment directed to me in the first minute of the first day by a faculty member who I had just met was, 'You must join our department basketball team.'" The fight against these stereotypes levied "an emotional tax that is a form of intellectual emasculation," he told the audience. "As of this afternoon, my Ph.D. will bring the national total of Black Astrophysicists from 6 to 7 (out of 4,000 nationwide). Given what I experienced, I am surprised there are that many."[28]

The implicit racial bias continued to be a part of his career. He later wrote: "Even astrophysicists have told me things they thought I should do. And I think it's because they could not picture me as their research colleague. Their equal." He added, "When I go to these astronomy conferences, there's nothing overt. Still, it's clear that when I get up there to talk, they're not used to having a Black person explain astrophysics to them."[29]

Over five years in the 1990s, Tyson estimated that he was interviewed on network television as an expert in cosmology more than fifty times. The decade saw a greater diversity of active black public intellectuals.[30] Yet the personal challenges and victories of black intellectuals were still presented in public as part of the larger development of black history. Henry Louis Gates Jr., for example, wrote that he felt an onus as a graduate student to examine black literature as if he had "embarked upon a mission for all black people." He wrote in the *New Yorker:* "it is the birthright of the black writer that his experiences, however personal, are automatically historical."[31]

Critical writing from a black perspective in the 1990s also positioned Tyson within the larger narrative of black history. As Bruce Caines said in *Our Common Ground: Portraits of Blacks Changing the Face of America* (1994): "Just by being a highly respected astrophysicist, he has made a statement that will outlive him."[32] *Strong Men Keep Coming: The Book of African-American Men* (1999) described Tyson as a "latter-day Benjamin Banneker," a description that draws a literal and symbolic continuity with the pioneering, self-taught eighteenth-century mathematician, astronomer, and inventor. Banneker designed and implemented a survey of the District of Columbia in 1790 and wrote to Thomas Jefferson, then secretary of state, challenging the contemporaneous strain of thought that blacks were mentally inferior.[33] Even so, Tyson is aware that he carries a burden of representing the black community,[34] though he has strategically avoided speaking during Black History Month "on the premise that my expertise is neither seasonal nor occasional."[35]

BRINGING ASTROPHYSICS TO THE PUBLIC

By the middle of the 1990s, Tyson's communication skills attracted the attention of the scientific elite. Ellen V. Futter, president of the American Museum of Natural History, told *Scientific American*: "He came into our field of vision as someone who was extraordinarily talented as a communicator of science."[36] In 1994, he was appointed staff scientist at an institution that was part of the larger organizational umbrella of the American Museum of Natural History—the Hayden Planetarium in New York.

The same year, he also began writing a monthly column called Universe in *Natural History*, a magazine hosted and funded by the museum that aims to promote public appreciation of the natural world among its college-educated and civic-minded readership, which numbers between fifty and one hundred fifty thousand.[37] In 1995, Tyson became the planetarium's acting director.

Tyson's *Natural History* columns combine the presentation of scientific facts with explanations of how science operates as a body of knowledge. He has discussed astrophysical topics, such as universal constants, cosmic microwave radiation, and the impact of light pollution on cosmology research. He has explained aspects of the scientific method, such as the centrality of uncertainty in science, how bafflement drives scientific curiosity, and new ways that nature can be examined through technology. Tyson also uses the columns to extend his intellectual range and reach. He addresses aspects of science and culture that have recurred over the course of his career: the inaccuracy of movies' portrayals of science, the wonder engendered by the cosmos, the benefits of the space program, and reflections on the relationship between science and religion.[38]

Tyson first articulated in his columns his idea of the cosmic perspective. Essentially a way of using cosmological ideas about the scale of history to give a more proportionate perspective on human life, the cosmic perspective "flows from fundamental knowledge," "comes from the frontiers of science," "opens out minds to extraordinary ideas," "enables us to see beyond our circumstances" on Earth and "not only embraces our genetic kinship with all life on Earth but also values our chemical kinship with any yet-to-be discovered life."[39]

Tyson's institutional affiliation with the American Museum of Natural History deepened in 1996 when he was appointed the inaugural Frederick P. Rose Director of the planetarium.[40] The position gave him a major platform as a scientist in public, as the planetarium described its aims, on its home page and in the message to the public from Tyson, to "bring the frontier of astrophysics" to wide audiences, as it has historically "blended scientific scholarship with innovative public outreach." His first major project was the management of the $230 million redesign of the museum's older planetarium, which was re-created as an eighty-seven-foot-wide sphere suspended inside a seven-story, ninety-five-foot-tall glass cube, on the western edge of Central Park.

Tyson was positioned as the planetarium's public face. On New Year's Eve in 1999, a black-tie fundraiser for the planetarium was promoted with a profile in the *New York Times* of Tyson in its "Public Lives" section. The article merged Tyson's public and private lives. It called him "an astrophysicist and bon vivant" with a "[b]ooming laugh" who has on his desk an "iron-nickel meteorite that fell to Earth 4.6 billion years ago." The piece portrayed Tyson as a refined scholar who collects "centuries-old books about contemporaneous scientific marvels and predictions about the future" and who writes in "reflective moments at home in downtown Manhattan . . . by candlelight with a quill pen."[41]

The newspaper associated Tyson with Carl Sagan. It reported that guests at the planetarium's fundraiser viewed a show that carried them through a five-billion light-year tour of the cosmos. The paper quoted Ann Druyan, who cowrote the show and is Sagan's widow, as saying that her late husband "would have loved this evening's voyage, and he inspired it." The journalist also pointed to Tyson's awareness of the value of publicity for the planetarium: "Like his idol, Dr. Sagan, he has a giant-star ego and media-savvy timing. Since this interview will appear on a holiday, he wonders, won't circulation be lower?" The profile ended by explicitly linking Tyson to the planetarium: "Tonight . . . he will be partying at the Rose Center, where his future, as well as that of the planetarium, seems luminous."[42]

The official opening of the redesigned planetarium in 2000 featured an orchestrated communications campaign to promote Tyson as a public personality. A variety of media profiled him that year. *Ebony* magazine, where the African American elite have historically showcased their achievements, featured a range of themes in its profile of Tyson. It noted his historical significance as it described him as "the first African-American since Benjamin Banneker to produce a book on astronomy." It stressed Tyson's role as a representative figure for minority students who wished to become scientists.[43] Tyson told the magazine: "My hope is that by being visible, I'm opening doors so that people better than I am, who may never even have thought they could do this, will come along and be revealed to modern society."[44] It noted his suitability for television, writing that with "his thick, helmet-like Afro, his quirky star-strewn ties and vests, and his infectious enthusiasm for all-things cosmic, Tyson . . . cuts quite a telegenic figure." It stressed his physicality and called him "a strapping ex-wrestler and Afro-Caribbean dancer." A photograph that accompanied the article showed Tyson in his apartment with his wife and daughter, illuminating further his private world.

People named him its "sexiest astrophysicist." It said: "the 6'2" Tyson indulges his love of wine and gourmet cooking while succumbing to the gravitational pull of his wife of 12 years, mathematical physics Ph.D. Alice Young, 44, who is expecting their second child next month."[45] *Wine Spectator* magazine described his wine collection and detailed how, in the 1980s, Tyson "developed a fondness for Gran Reservas from Rioja and systematically assembled a fourteen-vintage vertical collection of Dunn Cabernet Sauvignon Howell Mountain." But his schedule does not allow him to entertain as often as he would like—"he uncorks only two to four bottles a week."[46]

In 2000, Tyson also published the first edition of his autobiography, *The Sky Is Not the Limit*. He noted in the book's preface that he was not a film or sports star or important political figure, whose lives were suited to being recounted in memoirs. He presented himself instead as "just a scientist—an astrophysicist—who has tried to bring the universe down to Earth for everybody who wanted to see it." He described himself as an educator who "tried to raise public literacy in science." He framed his life history as one in which he overcame racial obstacles as he moved "against the winds of society" through "traumatic moments where my will, my life's goals, and my sense of identity were tested to their limits."[47] The *New York Times* in its review said the book was strongest when Tyson connected his personal life history to his career development.[48]

As well as discussing his life and career within racial contexts, Tyson used the autobiography to create associations with Sagan. He described how he met the famous physicist while he was choosing a physics department for his undergraduate study. Sagan showed him around his laboratory at Cornell University, gave him a ride to the bus station and, because it was a snowy evening, gave him his home phone number in case he was stranded in Ithaca, New York. Tyson wrote: "I never told him this before he died, but at every stage of my scientific career that followed, I have modeled my encounters with students on my first encounter with Carl."[49]

PLUTO KILLER

Tyson was at the center of a landmark decision in space science—that Pluto was no longer a planet. As part of its redesign, the planetarium changed the presentation of the solar system. It organized celestial bodies not into planets and moons—but into categories of similar objects. It placed Pluto among the increasing number of icy objects located in the Kuiper Belt, a vast field beyond Neptune densely populated with celestial objects and found by astronomers in the 1990s. Pluto's status was now ambiguous; it was not included in the presentation of planets, but it was not labeled as lacking planetary membership. The planetarium—with this presentation of the solar system—supported the effective demotion of Pluto.

Before the redesign, Tyson had stated his position on Pluto. In a 1999 *Natural History* column, he said that as a citizen, he wanted to defend Pluto's status as a planet—discovered in 1930 by Clyde Tombaugh at an observatory in Arizona and tied in popular consciousness to the Disney character.[50] But as a scientist, he had to vote for demotion.[51] The controversy played out largely among astronomers and interested citizens—until 2001. That January, the *New York Times* published a front-page story by journalist Kenneth Chang—headlined "Pluto's Not a Planet? Only in New York"—that said Pluto had been ejected from "the pantheon of planets."[52] The article had an instantaneous impact. Chang recollected in 2009: "I knew I had written a good article. I thought it was one of those 'fun reads' that people would enjoy, share with friends, and forget in a week. Man was I wrong. I turned Dr. Tyson into Public Enemy Astronomer #1, for years."[53]

The decision to categorize Pluto as a Kuiper Belt object was taken by the planetarium's scientific committee, but Tyson noted that he "became the most visible exponent of the decision."[54] Astronomers who wanted to maintain Pluto's planetary status used his public profile as evidence of a lack of credibility. One felt Tyson was "full of baloney," as he had not the required expertise in planetary geology to contribute to the debate.[55] "Is the problem perhaps that the Pluto controversy has been stirred up by a planetarium," asked another, "given that many professional astronomers are still inherently prejudiced against anyone who deigns to dedicate their time to the popularization of astronomy?"[56] In response, Tyson said the planetarium's motivation was not an attempt to generate publicity or cause controversy, but was educational.[57]

Tyson became not only a focus for specialist debate. He became a figurehead for public debate about central issues in planetary science: definition, categorization, and taxonomy. He also was a figurehead for the wider cultural issues associated with Pluto's position. In *The Pluto Files* (2009), his account of the controversy, he wrote: "Pluto's demotion became a window on who and what we are as a culture, blending themes drawn from party politics, social protest, celebrity worship, economic indicators, academic dogma, education policy, social bigotry, and jingoism."[58] One science studies scholar argued that the public discussion of these issues, centered on Tyson, showed the influence that public scientists at prominent institutions had on the inner workings of science.[59]

Pluto's status was eventually settled in 2006 when the International Astronomical Union defined planets in a way that excluded Pluto. Tyson had a high profile within popular science before Pluto's demotion, but Chang said the controversy helped bring him to a wider national audience. "The controversy raised his stature," said Chang. "He had a career before this, but I joke that, without me, he would be a two-bit planetarium director."[60] Former *USA Today* science writer Dan Vergano said the controversy established Tyson as "a personality."[61]

He also developed direct political influence. President George W. Bush appointed Tyson between 2000 and 2004 to two aerospace-focused committees. The government appointments for Tyson—who once identified his political values as "left of liberal"[62]—came during political debates about the use of expertise and scientific evidence during the Bush administration. Liberals, environmentalists, and some scientists argued that Bush and Republican leaders in Congress fostered a climate of "anti-science" and undertook a "war on science" in which they valued political ideology over expertise, especially in the areas of climate change, stem cell research, and environmental regulation.[63] Tyson objected to the "stereotype" that Bush was hostile to science: the broad portfolio of science as a whole, including physics and health research, benefited from increased funding under Bush's presidency.

THE *NOVA* MAN

A new edition of Tyson's autobiography was published in 2004. Its revised introduction showed that Tyson had an enhanced awareness of his own symbolic value. No longer did he present himself as just a scientist. He was

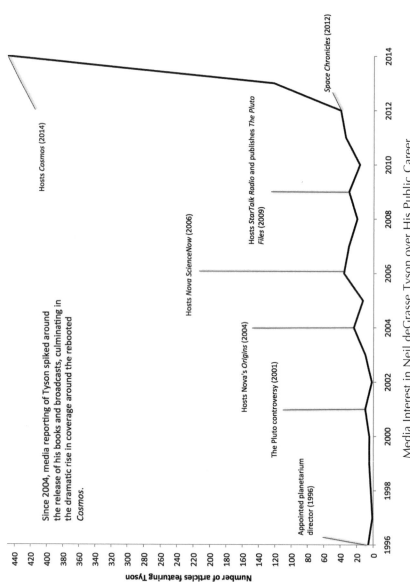

Media Interest in Neil deGrasse Tyson over His Public Career

Number of articles featuring Tyson

Appointed planetarium director (1996)

The Pluto controversy (2001)

Hosts Nova's *Origins* (2004)

Hosts *Nova ScienceNow* (2006)

Hosts *StarTalk Radio* and publishes *The Pluto Files* (2009)

Space Chronicles (2012)

Hosts *Cosmos* (2014)

Since 2004, media reporting of Tyson spiked around the release of his books and broadcasts, culminating in the dramatic rise in coverage around the rebooted *Cosmos*.

a unique scientist, who was aware that his combined experiences as a nerdy kid, college athlete, astrophysicist, and black man offered "what may be an uncommon portal through which to view life, society, and the universe."[64] This individual distinctiveness, coupled with his communication skills, meant he had the potential to perform through the medium that could bring his public intellectual work to the widest possible audience: television.

Tyson joined with another established cultural institution that had a long-running commitment to increasing scientific literacy: Public Broadcasting Service's *Nova*. The show's aim is to promote science to wide audiences, and it notes on its webpage that it draws a weekly average of more than six million viewers.[65] Tyson fronted the critically acclaimed two-part series "Origins" about the evolution of the cosmos. The *New York Times* noted that the show functioned also as a promotional vehicle for Tyson, designed to "send the young astrophysicist into orbit as a space-savvy celebrity."[66]

The promotion worked. From 2004, as the chart shows, the media attention to Tyson spiked when he published a book or broadcast a TV show.[67] From that year, he became a feature on the cultural landscape.

The *Montreal Gazette* in its preview of the show said: "Tyson tries to do for astronomy on television what Carl Sagan did more than two decades earlier with 'Cosmos': namely, answer the eternal question of our place in the universe."[68] *Publishers Weekly* said the show was "the most informative, congenial and accessible general look at cosmology to come along since Carl Sagan's 'Cosmos' 27 years ago."[69] The *New York Times* also wrote: "A quarter-century after Carl Sagan's 'Cosmos' series, television's astro-impresario is Neil deGrasse Tyson."[70] The book *Origins: Fourteen Billion Years of Cosmic Evolution* (2004) was published at the same time as the show was broadcast.

His television work brought him to wider public notice. "Origins" also proved that Tyson could carry a big-budget production, and in 2006 he was presenter of the second season of *Nova ScienceNow*, a short magazine-format science news program that is an offshoot of the traditional hour long *Nova*.

The promotion of the show framed Tyson as a distinctive personality. A PBS press release called him "one of the world's most popular lecturers on astronomy." The show's senior executive producer, Paula Apsell, stressed the combination of Tyson's credentialed expertise and unique personality made him a captivating host. "Neil's scientific background and the passion he brings to his work complement perfectly the series' commitment to reporting the most astonishing stories from the frontlines of science-in-process," she was quoted as saying. "He loves a good story—and it shows. His enthusiasm is palpable and infectious—a winning combination."[71]

Tyson exemplified the wider trend in television toward "presenter-led" programs. This genre of television featured personalities already known to audiences *before* they hosted shows. Consequently, these hosts had a dual promotional function: they brought an existing fan base to the show and exposed the show to wider potential audiences.[72] Tyson embraced his role as host and proved willing to occasionally ham it up to engage viewers. For example, the first show in the 2009 season portrayed Tyson as Indiana Jones, dressed as a 1920s police officer, and—to introduce a segment about voice enhancement—singing in a shower. The *Montreal Gazette* in a review of the series identified one of the show's strengths as "Tyson's unchecked enthusiasm and passion for his subject."[73] He said he wanted to make "science accessible and relevant to many different audiences."[74] Reporting on Tyson taking over the show, the *Daily News* said Tyson was "on a mission to make us 'science literate.'"[75]

Now that Tyson was an established media figure, journalists probed beneath his public image. The *New York Times* called him "a tall, dark and telegenic scientist with a booming voice . . . who delights in noting that he was born the same week that NASA was founded in 1958." The paper described how he was a "well-dressed and agile speaker," who thinks "science is fun," and who "plays against the nerdy scientist stereotype." The profile sought to illuminate Tyson's personality by describing his surroundings. It noted that one wall of his office was dominated by a print of van Gogh's *Starry Night*.[76] The *Washington Post* called him a "teddy-bearish guy who can be hammy at times, giggly at others" who "seems born for the role" of television presenter. The paper said Tyson was "magnetic in a way that suggests you can ask him anything and he will answer, even if the question is a bit idiotic."[77] Calling him the *Nova* man, the *Post* in another article called him a "[s]uper-slick astrophysicist/TV personality."[78]

In 2007, the cultural impact of Tyson was clear. *Time* named him among its one hundred most influential thinkers in the world. *Death by Black Hole*, a collection of Tyson's *Natural History* columns, debuted at number 14 on the *New York Times* best selling nonfiction list.[79] Comparisons with Sagan became routine. "You can think of Neil deGrasse Tyson as the Carl Sagan of the 21st century—as long as you envision a Sagan who's muscular, African-American and as cool as his predecessor was geeky," wrote Michael Lemonick in *Time*. "These minor differences aside, Tyson

. . . is the undisputed inheritor of his late predecessor's mantle as the great explainer of all things cosmic."[80]

He classified himself as an agnostic. The conflict between science and religion was restricted to schools, he said in the skeptical and rationalist-orientated *Point of Inquiry* podcast. "If you're going to tell me that Noah had dinosaurs on his ark, I'm sorry, you are ignorant and scientifically illiterate, and you don't belong in a science classroom."[81] Contributing to this issue, he helped create *Science, Evolution, and Creationism*, a 2008 report—sponsored by the National Academy of Sciences and Institutes of Medicine—aimed primarily at educators that outlined why public schools should only include scientific explanations of evolution, but stated that accepting evolution was compatible with religious faith.[82] Tyson separated himself from those he called "in-your-face atheist." In contrast, he said: "I'm a scientist. I'm an educator. My goal is to get people thinking straight in the first place. . . . That's what I'm about."[83]

Death by Black Holes also repeated another theme in his work: how inaccurate film representations of the universe harmed scientific literacy. *Titanic* (1997) featured characters dressed in the fashions of 1912, but did not present faithfully the constellations that were visible the night the ship sank. (Tyson told this to the film's director, James Cameron, in person and, two months later, the filmmaker presented a post-production specialist with an accurate sky to be used in the Special Collector's Edition of the film.)[84]

Science literacy was for Tyson a sort of magic bullet for social ills. For example, he told one interviewer that science literacy efforts should not only be aimed at children because it was "the adults that need the science literacy, the kind of literacy that can transform the nation practically overnight."[85] "When you're scientifically literate, the world looks different. Science provides a particular way of questioning what you see and hear. When empowered by this state of mind, objective realities matter," said Tyson when asked in 2011 by the *New York Times* what he would do as president. "These are the truths on which good governance should be based and which exist outside of particular belief systems. Our government doesn't work—not because we have dysfunctional politicians, but because we have dysfunctional voters," he said. "As a scientist and educator my goal wouldn't be to lead a dysfunctional electorate, but to bring an objective reality to the electorate so it could choose the right leaders in the first place."[86]

Tyson's talent for television meant he was able to communicate science in unconventional media shows. He was the first physicist to appear on the satirical news shows *The Daily Show* and *The Colbert Report*, shows that provide a gateway to science content for people who may not have a strong interest in science-related news. The more people watch these shows, the more attention they pay to science news.[87] Tyson became so adept at these particular media performances that he gave advice to other scientists asked to appear on late-night talk shows. Discussing his strategy for doing *The Tonight Show with Jay Leno*, Tyson said he studied the usual length of answers that guests gave before Leno interrupted with a joke. As a result, he said he parceled his messages into three-sentence chunks: "Soundbite it down," he advised. "Hand him a soundbite he can't edit."[88]

He was now so instantly recognizable that he appeared in other shows *as himself*. He made a cameo appearance as himself on the comedy show *The Big Bang Theory*, dressed in his signature moon-covered vest or waistcoat, being criticized by the character Dr. Sheldon Cooper for demoting Pluto.[89] Tyson appeared as himself in a 2008 episode of *Stargate Atlantis*[90] and was also a cartoon version of himself on the PBS children's show *Martha Speaks*. At the request of DC Comics, he appeared as himself in an edition of a Superman comic where he helps the hero glimpse his home planet, Krypton. [91]

Yet work as a public scientist took him away from his scientific work as a skilled astronomical observer. He has published scientifically on his specialist areas, with papers in leading journals such as the *Astronomical Journal*, but his research career has been modest. By June 2012, his *h-index*—a figure that demonstrates the impact of a scientist's research, the higher the better—was 9. In 2005, Hawking's was 62.[92]

FUNDING THE CELEBRITY SCIENTIST

Even so, the scientific establishment backed Tyson financially. He was the principal investigator on grants awarded by the National Science Foundation (NSF) to fund, in part, a novel science radio show, *StarTalk Radio*, which planned to combine astronomy and comedy to bring science to a diverse listenership. In 2009, the NSF awarded him $200,000 and the description of the grant stressed Tyson's star power, describing him as a popular best selling book author, host of PBS's *Nova ScienceNow*, and a frequent contributor on cable television networks, including *The Daily Show*. After the initial ten shows were broadcast, Tyson received another NSF grant in 2010 that totaled more than $1.2 million.[93]

StarTalk Radio wanted to reach listeners traditionally ignored by science media. First broadcast on AM radio, the show was aimed at what Tyson called the "blue-collar intellectual." It presented offbeat science items: the science of wine, the science of the science fiction show *Star Trek*, the physics of time travel. Morgan Freeman, Whoopi Goldberg, Jon Stewart, and actors and actresses from *Star Trek* were among its special guests. Its interview-based format allowed for a broad discussion of science.

For example, an interview with Nichelle Nichols, who in her role on *Star Trek* in the 1960s was the first African American character in a major television series who did not occupy the role of servant, addressed the representation of diversity in science fiction.[94] And Whoopi Goldberg discussed her desire to appear as a black character on *Star Trek* because she wanted to be an emblem of the diversity she felt was lacking in science fiction shows.

Goldberg turned the focus on to Tyson. She argued he had a pioneering role as an African American scientist. "Science is always sort of touted as something we won't understand, we won't get, normal people," said Goldberg. "It also was thought of as being something that was an elite thing that we mere mortals were not nearly smart enough to understand. And here you come, not only are you not just a mere mortal, but you're a big ol' black man." Goldberg placed Tyson alongside the similarly emblematic figures of Tiger Woods and Barack Obama. She said: "Believe me, when I was a kid, there were none that I was aware of, and so suddenly, like golf in a weird way and like the presidency, these things now have become possible."[95]

Tyson embraced Twitter. Followed by more than two million people in summer 2014, Tyson mainly tweeted cosmological facts and trivia (he is preparing a volume of them for publication).[96] But his tweets also maintain his celebrity profile, as they provide what media researchers call "backstage access": a view into his personal life that enhances intimacy between him and his fans. For example, on October 13, 2012, he wrote: "I've come to conclude that Fettuccini Alfredo is just Mac-and-cheese for food snobs." On July 25, 2012, he responded to a fan in more serious vein: "@adinasauce: My uncle passed tonite. Your thoughts on our atoms & the universe comforted me. Thank you. //No, thank the cosmos."

THE NASA ADVOCATE

Tyson fashioned himself in recent years as NASA's chief public advocate. The beleaguered space agency needed him after it lost its glamour since the pinnacle of its popularity and funding in the late 1960s. Its public image never recovered from the *Challenger* and *Columbia* disasters in 1986 and 2003, as well as the incorrect manufacture on the Hubble Space Telescope.[97] Another famous scientist of the twentieth century, Richard Feynman—famed within physics for creating new ways of thinking about quantum mechanics—came to wide public attention after he spoke about NASA's incompetence as part of the official investigation of the *Challenger* disaster.[98]

Tyson's support is an example of an established strategy by NASA to garner citizen and policymaker support: public visibility. The agency and its advocates traditionally used three arguments to construct a favorable public image. Nationalist arguments said the space program benefited the nation. Romantic appeals argued space travel fulfilled basic emotional yearnings for discovery and exploration. Pragmatic appeals showed how the space program led to tangible benefits in people's lives.[99] Tyson marshaled all those arguments in 2012 as he sought to revitalize NASA's public profile. The fall in the agency's fortunes, he argued, mirrored the fall in the nation's economic power. As NASA rises, he said, America rises.

He used his fame to press the case in a variety of venues. The argument for NASA's continued importance ran through the space-focused essays collected in *Space Chronicles* (2012), which spent a week on the *New York Times* best seller list. An excerpt of the book appeared in *Foreign Affairs*, putting its arguments directly before Congress.[100] Tyson told news agency Reuters: "If you plan a mission to Mars and select the new astronaut class, you'll change the attitude and mood of the country."[101] On *Late Night with Jimmy Fallon*, Tyson said, "NASA needs more money, period. Whatever then happens I'll be happy with."[102] He told *The Colbert Report*: "I'm telling you the manned [space] program is a force of nature on the educational pipeline of America. It is the force that excites people to want to become scientists in the first place, the fact that NASA can dream big about where to go next in space."[103]

He testified about NASA before Congress. "All these piecemeal symptoms that we see and feel—the nation is going broke, it's mired in debt, we don't have as many scientists, jobs are going overseas—are not isolated problems," he said in his testimony. "They're part of the absence of ambition that consumes you when you stop having dreams." He advocated doubling NASA's budget, which currently stands at half a penny from every tax dollar, an increase that he argued would lift the country out of its current depression.

NASA has traditionally been politically bipartisan. Tyson positioned himself also as bipartisan. He is careful to adopt neutral positions in science policy debates. When he campaigned for NASA, he took on what science policy experts call the role of the "issue advocate," who merely advocates for a policy, in this case more investment in space science. When he positioned himself within partisan policy debates, he took on the role of the "science arbiter" who answered factual questions from decision makers.[104]

For example, when comedian and left-leaning political activist Janeane Garofalo criticized him on *StarTalk Radio* for not questioning a Republican interviewee as aggressively as she wanted, Tyson said he was a "centrist advocate." He said: "I stand in the middle and look at the hot air on both sides. . . . Science is a stand that can be taken strongly in the middle. . . . I claim there's a place to stand in the center where everyone comes to you because you end up speaking the truth."[105]

Tyson also once visited the home of David Koch, the billionaire and conservative activist, who has given $22 million to the Museum of Natural History and sits on its board of trustees, and who supports politicians skeptical of climate change. Tyson taught members of his family how to work a telescope, something he did also for former New York mayor Michael Bloomberg.[106] *Vanity Fair* called him "a paradox of persuasive rationalist and romantic Space Age dreamer."[107]

REBOOTING *COSMOS*

In 2014, Tyson remade Sagan's famous show *Cosmos*. The rebooted *Cosmos* resulted from a cross-pollination of fame. Ann Druyan, a creator of the original and Sagan's widow, had long wanted to resurrect the iconic show for a new generation with Tyson as host. The idea gained traction after Tyson met Seth MacFarlane, the creator of the long-running Fox show *Family Guy* and a Sagan fan who had loved the original *Cosmos* as a child.

They were introduced by the Science & Entertainment Exchange, a National Academies of Science program that connects show business executives with scientific experts in an attempt to make the science in movies and TV shows accurate and exciting.[108] MacFarlane used his influence at Fox to advocate for the show. The network committed to thirteen episodes and, rather than aiming for a niche audience, as the creators of science shows usually do, drove *Cosmos* deep into public culture.[109]

Fox's investment was a commercial risk. The original was broadcast when television was a cornerstone of culture, a medium that drew wide audiences and sparked social conversation. But the new version came as television audiences dwindled and splintered among hundreds of channels that competed not just with each other but also with a vast array of other sources of information and entertainment, each tailored to narrow audience tastes. But Fox went wide: *Cosmos* was available overall on 220 channels in 181 countries in forty-five languages, targeting half a billion homes. About 8.5 million viewers watched the premiere of the show, the total rising to 40 million globally once you include international viewers and those who recorded it to watch later.[110]

Determined to match the original as a spectacular piece of landmark television, producers recruited Hollywood talent to give *Cosmos* Mark II first-rate production values. There were lavish visual effects. The Spaceship of the Imagination, a feature of the original, brought viewers through the icy rings of Saturn into the outer darkness of the cosmos and into the microscopic world of the cell. Tyson witnesses a visceral re-creation of the big bang, his suit jacket swept back as the universe expands from its mysterious origin in an instant of terrifying force and brightness.

Neil deGrasse Tyson, wearing his customary universe-themed vest, promotes *Cosmos: A Spacetime Odyssey* to journalists in 2014 ahead of the broadcast of the thirteen-part story of life and the universe. The show's first episode attracted an estimated forty million viewers worldwide. Photo by Frank Micelotta/Invision for FOX/AP images

The first episode portrayed Tyson as Sagan's successor. In the opening segment, he walked the same California coastal cliffs that Sagan strode in the original. At the end of the same episode, he opened Sagan's 1975 diary that showed an appointment he noted for Saturday, December 20. It said: "Neil Tyson." The *Washington Post* said Tyson was "a worthy heir to Sagan's legacy."[111] *Variety* said Tyson "evokes a sense of scientist-as-rock-star cool and lends authority, recognizability and social media cachet."[112] Tyson was at the forefront of promotion for the show. At a press event to discuss the show, Tyson wore his trademark starry vest as he conveyed some of the ideas covered in *Cosmos*.

But its producers saw it as more than just a beautifully shot bonanza of amazing science. "We've seen the rise of junk science, of charlatans who are telling us to not vaccinate our children . . . the rise of schools questioning evolution," said MacFarlane. "All these things piling up that betray the fact that we've lost our way in terms of our scientific literacy." He wanted *Cosmos* to get America's scientific understanding "back on track." Druyan was as forceful. "That would be amazing—to have a public that will support science," she said, "but also a public that will also think critically and be less subject to the forms of manipulation that we all know are happening commercially, spiritually, governmentally."[113]

The *New York Times* science writer Dennis Overbye hoped the show would provide a fundamental set of understandings that would inform public discussion of science controversies, such as fracking and genetically modified food. "If everybody watches the new 'Cosmos,'" he wrote, "we can talk about it the way we once argued about 'The Sopranos' every Monday morning."[114] Yet other writers doubted a television show could alone transform scientific illiteracy into scientific appreciation, especially as the original was broadcast at a historical moment when tensions between the United States and the Soviets simmered. "Public support for science was an easy sell, in part," wrote historian of science Audra Wolfe, "because so much of the Cold War rivalry depended on high-tech weaponry built on cutting-edge science."[115]

The rebooted *Cosmos* ignited simmering cultural tensions in U.S. public life. The fact of human-induced climate change was beyond doubt, Tyson stated, and the evidence for evolution was uncontestable. By taking a position on these politically polarized issues, he became a target for conservative critics. *National Review* columnist Charles C. W. Cooke called the astrophysicist "the fetish and totem of the extraordinarily puffed-up 'nerd' culture that has of late started to bloom across the United States." The term "nerd," Cooke argued, has come to signify someone with liberal politics.

Tyson, therefore, the columnist continued, became a symbol, "deployed as a cudgel and an emblem in argument—pointed to as the sort of person who wouldn't vote for [Republican] Ted Cruz."[116]

More damaging was a series of articles published in the right-leaning online magazine *The Federalist* that accused Tyson of making up quotes and stories. The most prominent anecdote scrutinized by the magazine concerned Tyson's claim that, after the 9/11 attacks, President Bush said, "Our God is the God who named the stars," a pronouncement the scientist interpreted as the president's attempt to segregate Christianity and Judaism from fundamentalist Islam. Bush's speechwriters denied the president said it. Tyson said he remembered it. But the physicist later conceded that he had mistakenly taken a quote Bush made in the aftermath of the tragic Columbia space shuttle crash in 2003—"The same Creator who names the stars also knows the names of the seven souls we mourn today"—and transposed it to the traumatic days following 9/11.

Summarizing the minor scandal, law professor and contributing editor to the *National Review* Jonathan H. Adler wrote in the *Washington Post*: "The evidence here clearly shows Tyson screwed up." But he noted: "I am sure some of Tyson's political adversaries would like to use this episode as a basis for attacking climate science or evolution. No dice. It does not work that way."[117]

SCIENCE'S CELEBRITY DIPLOMAT

Tyson earned his scientific credentials, but his research career is middling. Unlike Hawking, he made no signature contribution. Unlike Dawkins, he did not crystallize in a major popular book a new way of understanding nature. Unlike Pinker, he does not glide between public and peer influence.

His fame rests in part on his rare talent for communication, a skill he relentlessly honed over time as he passionately engaged and entertained broad swaths of the general public about science. The talent emerged as he wrote pseudonymously as Merlin, explained space science to television journalists, wrote essays on astronomy, hosted a succession of television shows, and stands at the head of one of the world's major planetariums.

But his mastery of the arduous art of explaining science does not alone explain his fame. Neither does the significant fact that for decades the scientific establishment invested support in Tyson, providing him with a series of public platforms—from *Natural History* magazine to the Hayden Planetarium itself—that allowed him to exercise his ability to engage wide audiences.

He came to represent three wider historical movements in the tumult of post-1960s U.S. culture. He exemplifies, first, the rise of the African American public intellectual who contributes his expertise to a civic issue unrelated to race. Yet at the same time, he has come to represent the struggle of the black intellectual to reach an influential position in public life. In his case, he fought to achieve his status in science, with its history of racism, and in U.S. space science, with its own record of segregation.

Tyson, second, became the chief public advocate for greater scientific literacy, in part to recapture an imagined—though false—glorious era of public scientific understanding in the post-Sputnik drive to conquer space. Relatedly, Tyson has come to represent the drive to restore American public and political commitment to the type of space exploration exemplified in the Apollo and Mercury eras, an endeavor that would catalyze national pride and productivity.

In addition, as his career progressed, Tyson cleverly positioned himself as the successor to Carl Sagan, a notion journalists uncritically amplified. Tyson—though unique—stepped into the empty space on the cultural landscape for a scientist who could engagingly explain the cosmos to mass audiences. Plus, he was media friendly, not only in his ability to package sound bites, but in his crafted personal appearance: why else would he wear the galaxy-themed vests?

But what Tyson illustrates most strikingly is the new power of scientific celebrity. Tyson's media-driven celebrity has ultimately earned him social power and influence not only over nonspecialist publics, but also over science policy, the U.S. space program, the categorization of the planets, and the future of his scientific field. His stardom has allowed him to glide between these realms.

Publicly bipartisan, largely uncontroversial, media-friendly, and associated with a philosophy that sees science as essentially neutral in politics, so far in his career Tyson has been an engaging public face of U.S. science. In public, he is science's perfect diplomat. It is a precarious position, however, and a planned future book could attract controversy because its subject matter deals with one of the more contentious aspects of American research: the links between science and the military.

TEN

A New Scientific Elite

In January 1997, Stephen Jay Gould wrote an obituary in *Science* for his dear friend Carl Sagan. He praised the host of the original *Cosmos* as "a fine scientist and the greatest science popularizer of the twentieth century, if not of all time." But Gould also berated many elite scientists as hypocrites. They made what he called hollow pronouncements about the need to communicate science even as they downgraded the reputation of scientists, like Sagan, who spoke to broad audiences.[1]

But the view Gould expressed in the obituary was, by then, almost obsolete. The proof was his own career. Over more than four decades, he occupied a series of connected roles as talented paleontologist, revolutionary evolutionist, Harvard professor, captivating essayist, prolific author, antinuclear campaigner, and—for a generation of Americans—the public face of science. Three years after he wrote Sagan's obituary, the American Association for the Advancement of Science (AAAS) made Gould its president. That same year, the U.S. Congress named him a Living Legend.

Gould and the other scholars profiled in these pages became influential leaders in a new era of science, one entwined with modern media, one that no longer views popularizers as second-class scientists, one where scientists have enhanced responsibilities to speak to concerned citizens about the roles of science in society. They are the vanguard of a scientific community that needs

CHAPTER TEN

a strong public voice as it clamors for attention in a crowded public arena splintered into thousands of websites and hundreds of television channels.

These new public scientists are entwined with a defining feature of our media age: celebrity. Without celebrity, scientists would no longer have a prominent place in popular culture. When scientists are celebrities, they give science a face, force, and an impact in public life. But the shift to celebrity they represent does not constitute a dumbing-down of science. Nor can their success be reduced to spin and salesmanship. They are more than just the best PR people in modern science. They demonstrate that it is through celebrity that ideas and issues are spread, encountered, and understood in our personality-focused cultural mainstream.

Star scientists, therefore, wield influence across public life. They have the prominence to make direct appeals to policymakers for funds and to contribute informed knowledge to science-related policy debates. They persuade an anxious public that science deserves its public funds because it improves individual lives, enriches communities, and solves social problems.

They also defend science from perceived public attacks, standing up for its cognitive power and record of accomplishment. And not least, they circulate scientific ideas through culture to enhance public understanding and to help science take its recognized cultural place alongside art and literature as one of the greatest feats of humankind.

But as well as influencing public life in these manifold ways, their celebrity is a source of authority within science, allowing them to affect the internal workings of scientific practice. As a result, the star scientists are what sociologists call boundary spanners, rare figures who can operate *at the same time* within the professional culture of science and public culture. As these worlds become more enmeshed, the star scientists, by working in both arenas, become more influential.[2]

They become a new scientific elite. They are essential to the scientific enterprise. They are vital to public life.

HOW CELEBRITY SCIENTISTS ARE FORGED

Not every scientist can join this influential collective. Even so, the scientific community, understanding the power of celebrity to communicate ideas, has somewhat clumsily tried to manufacture fame. The well-intentioned Rock Stars of Science advertising campaign, for example, pictured renowned re-

204

search scientists alongside musicians like Debbie Harry and Seal in the hope that some of the rock star glamour will magically rub off on the scientists.

But fame, real fame, lasting fame, cannot be manufactured or constructed or created. The public lives of the scientists profiled here show that they all became famous as two major cultural processes connected and overlapped: the process of developing into a public intellectual and the process of being transformed into a celebrity.

To become a public intellectual, each scientist first developed advanced expertise in their specialist fields. Gould had *Cerion*, for example, and Greenfield had her AChE protein. All established their scientific credibility. There is no scientific equivalent to the vacuous reality television star who is famous for being famous.

Second, they had available channels of communication to wider audiences. Dawkins, Pinker, Greene, and Hawking wrote books. (Although Hawking had already been the subject of significant journalistic attention, the book was exclusively his ideas.) Gould wrote for *Natural History*. Tyson and Greenfield were brought into the public eye through television. Lovelock first explained Gaia to nonscientists in *New Scientist* before writing *Gaia*.

Third, all the scientists voiced views that engaged over time with the concerns of broad audiences. The views of Dawkins, for example, on the explanatory power of Darwinism, the social role of science as arbiter of truth, and the lack of empirical evidence for religious belief engaged with—and articulated—the beliefs of broad audiences over time.

And last, all the intellectuals developed a reputation for voicing important views. Greenfield's opinions on digital culture drew vociferous criticism, but the views addressed a growing social concern, something illustrated by the amount of coverage her views generated.

When the scientists first communicated with broader audiences, the process of celebrification started. The scientists themselves and journalists merged their public and private lives. Their words and images also became valuable to publishers and broadcasters: the scientists became cultural commodities. The Hawking franchise has been going strong for almost three decades: *A Brief History of Time*, *A Brief History of Time* (movie), *The Illustrated A Brief History of Time*, *A Briefer History of Time*, *My Brief History*. The Lovelock franchise has been running for four decades: *Gaia*, *Ages of Gaia*, *The Revenge of Gaia*, and *The Vanishing Face of Gaia*. Dawkins fans can buy from his website *An Appetite for Wonder*, "The Enemies of Reason," and "The God Delusion" T-Shirt.

But the last part of the process of making a celebrity is the most elusive, most abstract, yet most important feature of fame. As cultural critic Louis Menand explained, this is the way a star's personality intersects with history, the way his or her stardom coincides with the Zeitgeist, the spirit of the age, so that there is a perfect correspondence between "the way the world happens to be and the way the star is."[3]

Tyson thus embodied and reflected the post-1960s struggle of African American intellectuals. Lovelock represented the growth and influence of the broad church of ecology as the world struggled with the threat of climate change. Greene's work on the strange worlds of string theory became the face of twenty-first-century physics. This elusive ability to embody social trends describes why someone like the much-loved Bill Nye (the Science Guy) is a valuable and well-known public figure in the United States but is not a lasting celebrity comparable, say, to Stephen Jay Gould. Nye is primarily a translator of scientific ideas rather than a figure whose own image or ideas have had an impact in popular culture.

The star scientists not only represent and reflect particular historical moments. They also offer explanations of how the world works—explanations that meet a deep cultural need among the broad public. They articulate and express the views of a wide section of the general public. The dial of public interest, in the words of Menand again, "seems permanently tuned to the frequency at which the individual star is broadcasting."

As the twenty-first century dawned, audiences agreed with their views of the world or found them persuasive or sympathetic. It was a time when a plurality of belief systems—religious, spiritual, philosophical, nationalistic, all claiming to make sense of the world—collided in public. An unacknowledged consequence of postmodernism was the difficulty in determining true, reliable knowledge of the world. So many expert views clashed in the media, making it difficult for the nonspecialist to discern where the balance of evidence stood on any issue.

In this flux, science became not just a collection of methods for determining verifiable facts of nature, but, through the star scientists' writings, became a set of ideas for understanding culture, society, and human life. Their views of science offered ways of understanding the world that provided a clear path through the morass of contradictory and fragmented knowledge in the modern era. They provided reliable knowledge, clear meaning, and systemic understanding. Hawking, Dawkins, and the others became stars because they were just the figures the public wanted. They answered a cultural yearning.

Pinker, for example, in his books, advocated a liberal, progressive view of society, one driven and underpinned strongly by biological explanations for human behavior, explaining through natural selection not only how we develop our bodies, but how we think and speak. Writing about Pinker's *The Blank Slate*, philosopher of science Simon Blackburn said the book sold so well because it offered the "promise of a new synthesis, a science of the mind that finally tells us who we are, what is possible for us, how our politics should be organized, how people should be brought up, what to expect of ethics—in short, how to live."[4]

That insightful idea applies beyond Pinker; it helps explain the public appeal of several other scientists. Dawkins offers a similar well-worked-out philosophy of life. He has consistently argued that human existence is driven by our genes' urge to survive, the truth of the natural world can be understood and explained by science, but the world is no less beautiful for that, and its evidence-based explanations cannot allow for the existence of God to explain reality.

Gould explained a world where a myriad of natural forces created a world of almost infinite and unique variety, a world where natural processes met accidents of history to explain the shape of a panda's thumb, but also DiMaggio's fifty-six consecutive hits. He also showed how science was not sealed off from the inequalities of the world, and how science often perpetuated those inequalities.

As public intellectuals, they *stood for* these views. As celebrity public intellectuals, they *embodied* these views.

At a more general level, together they personified the more abstract and universal concepts of reason, rationality, and truth. They make these ideas—foundational notions of Western civilization since the Enlightenment—real and recognizable. In celebrity culture, they are science incarnate.

REVEALING HOW SCIENCE
WORKS AND *REALLY WORKS*

With their star status, their ability to *stand for* ideas, their ability to make these ideas *real* and *vivid* and *personal*, these scientists wield influence in many ways in celebrity culture.

They, first, spread scientific ideas through culture, carrying specialized ideas to communities far beyond science. The first popularization of the new genetics was *The Selfish Gene*, a book that has sold more than one million copies and has become a modern literary classic. Hawking boasts that one copy of *A Brief History of Time* exists for every seven hundred fifty people on Earth. The book gave an impressionistic understanding of the edges of cosmological research. And its commercial success led to an explosion of interest from publishers and audiences in popular science.

The revolution in linguistics only rippled across culture after *The Language Instinct*. The intricacies of evolution were brought to a wide audience by Gould in his books and three hundred *Natural History* columns. The rebooted *Cosmos* was watched on its first night by more than eight million viewers. *The Elegant Universe* became a best seller with its dense explanations of esoteric physics.

Their books and broadcasts have led to discussions of their ideas in media. As the charts of media attention in the chapters showed, coverage usually peaks around the publication of books or the broadcast of their shows. At these times, the scientists discuss their science in interviews, or reviewers discuss the books, and news articles draw attention to the new ideas. The release of their cultural products guarantees a public discussion of their ideas. With their works, these stars soak culture in science.

Moreover, their ideas diffuse through culture in odd ways. As well as his books, Greene discussed his ideas on *The Big Bang Theory*, acted as a science consultant in the film *Frequency*, starred in the same film as a (sort of) fictional version of himself, and wrote a children's book that riffed on the myth of Icarus to introduce ideas about space and time. Greenfield's ideas on the dangers of digital culture spread through parliamentary questions, newspaper op-eds, and a novel. Celebrity helps bring science out of the science section of bookstores or the science and books pages of elite newspapers.

Because their ideas have spread through society, the star scientists have enhanced scientific literacy. Their books fizz with facts, theories, and concepts. Described in their pages are event horizons, singularities, braneworld scenarios, phenotypic effects, punctuated equilibrium, Calabi-Yau manifolds, deep structure, natural selection, climate feedback loops, strings, dark matter, dwarf planets, reconfigured synapses, and many other concepts. Even if audiences do not grasp all the technical points, the books offer an impressionistic understanding of these ideas, a sense of encountering science.

The celebrity scientists show us how science works. They illustrate the process of science by describing the workings of their own scientific minds. Hawking tells us how he visualizes cosmological concepts as he works out his equations. Lovelock tells us how he uses a technique close to empathy as, in one example, he imagined himself inside a freezing cell to understand what happens as it grows colder. Gould tells us how he solved scientific problems, not by a series of logical steps, but as a moment of understanding when several different pieces of the puzzle integrated in a plausible solution. Pinker shows us his style of science as he expressed his frustration when he first encountered the vague and mushy theories of language learning that so contrasted with the clear, law-like concepts he used in his studies of visual cognition.

More broadly, these scientists show us the processes of how scientific fields operate. We see in *The Language Instinct* how cognitive science led to a radical new way of understanding how we come to speak. *The Blind Watchmaker* showed how evolutionary theorists reconstructed an unassailable case for evolution from a multitude of different sources of evidence from fossils to genes. *The Elegant Universe* showed how the laws of reality are expressed in advanced mathematics and how physicists judge the worth of ideas by supposedly subjective ideas of beauty and elegance. The progression of Gaia shows the strenuous experimental tests that ideas are subjected to before they can be counted as valid scientific knowledge—and even then, the ideas can be examined and overturned. *Wonderful Life* shows how years of painstaking classification and interpretation of fossils led to a new understanding of evolution, the chance happenings that resulted in the development of our earliest ancestors. They show us how science progresses.

The celebrity scientists also convey the joy of science. They show us what it *feels* like to be a scientist. Gould tells of his excitement of making a new discovery about snails, a finding that only interested about seven other experts, but brought him the unequaled pleasure of finding something new about nature, something that would forever be *his* finding.

Pinker tells us his study of irregular verbs brought him the deep satisfaction of knowing one thing well. Greene recalled the flash of joy as he realized abstract mathematic equations described the real-life motion of a bouncing ball. Tyson tells us that studying the universe gives him a connection to the cosmos that evoked what he considered to be spiritual feelings. Good scientists feel science in their bones. Their popular work allows us to feel it, too.

At a deeper and more important cultural level, the star scientists show us how science *really works*. They demonstrate, first of all, that there is not just one single thing called Science. Nor is there one single-by-the-numbers approach that can be called The Scientific Method. Reading their work illustrates the notion from science studies scholar Evelyn Fox Keller that "different collections of facts, different focal points of scientific attention, but also different organizations of knowledge, different interpretations of the world, are both possible and consistent with what we call science."[5]

Science is Lovelock taking air samples off the coast of Ireland and creating a computer model to show the evolution of Gaia. Science is Dawkins making meticulous observations of baby chicks in a laboratory and sitting at his writing desk to create a new way of seeing evolution. Science is Gould studying the shape of a tiny snail's shell and looking again at the entire fossil record to revolutionize paleontology.

Different fields have different ways of understanding. Take evolution. To read Dawkins is to see natural selection driven by the gene. To read Gould is to see a wider set of processes at work, such as mass extinctions, and that individuals and species, in addition to genes, drive evolution. Gould also relies on a different type of evidence, showing that the fossil record provides valuable insights on evolution. To read the volumes of writing by them and others about evolution is to see Gould's views as in the minority and Dawkins's views being critiqued by new movements in the field. To read some of Pinker's works is to see how ideas about evolution can explain human behaviors. Different forms of evidence, different nuanced intepretations, different views on how far these ideas can apply to the psychological and social worlds are all consistent with evolution. Evolution is true, the fundamental process shaping all life on Earth, but all its workings and finer points have not been settled.

And take physics. To read Tyson on astronomy is to read about cosmic phenomena that rely on a solid base of empirical evidence. To read Hawking on cosmology is to understand features of the universe, like Hawking radiation, that are accepted as real by scientists, but still lack empirical evidence. To read Hawking, too, is to understand how new instruments led to an explosion of interest and a newfound respect for cosmology as a field in the 1960s. To read Greene is to go a step further, to see the frontier of physics as a set of ideas about multiple realities that so far have no solid, accepted base in evidence. To read Greene and his detractors is to understand also the creation of a critical mass of like-minded researchers is essential to the advancement of knowledge. All are physicists, broadly conceived, but each presents varied topics, approaches, and evidence. It's all physics.

The star scientists bring into the public arena something all scientists know, but don't often discuss. Their work is filled with jealousies, rivalries, passion, personal visions, and fights for status. The public spats between

Dawkins and Gould and their supporters turned ugly (evolutionary biologist John Maynard Smith once said Gould was "a man whose ideas are so confused as to be hardly worth bothering with").[6] Lovelock was dismissed as just a guy who made things in his garage. The anonymous attacks on Susan Greenfield after she was rejected as a Royal Society fellow show that scientists can be vicious and vindictive.

The human dimension to science can also be seen in the accounts of their careers. The lives of Lovelock and Gould show that the scientific vocation is pursued through the gamut of joys and challenges that feature in all walks of life: raising children, divorce, illness, professional setbacks. The myth of the pure knowledge seeker does little to show the real nature of science. Einstein's reputation was protected until long after his death. In 1993, *The Private Lives of Albert Einstein* revealed him to be a serial philanderer who never saw his daughter and married his cousin. The book, for the late novelist J. G Ballard, was "a hand grenade lobbed into the sacred temple."[7] Despite the often ludicrous coverage of Hawking as a totally cerebral being, the intense media portrayal of him does not allow for his reputation to be protected and mythologized to quite the same extent. A close examination of his media portrayal over time shows him to be a man with a regular, messy, complicated private life. Science aims to touch the stars, but its practitioners are all in the earthly gutter.

The celebrity scientists show us facts and concepts, but they go further to show us how science works and how science *really works*. Celebrity culture melds argument and autobiography. It fuses private character and public science.

As a result, we understand the real nature of science through the stars' *public personalities*. That makes them crucial characters in modern public life.

SHAPING THE INNER WORKINGS OF SCIENCE

The celebrity scientists also have an enhanced power *within* science. It is not that famous scientists are somehow better able to produce new knowledge (although visibility has always led to prestige and the grants needed to do the day-to-day work of science). Nor is it that fame has replaced expert peer review of data, the traditional means of quality control within science. Instead it is that celebrity—largely bestowed on figures by the demands and interest of us general readers—is now an influence on what topics scientists research, how technical problems are defined, how ideas are understood, what areas are deemed worthy of investment, and how the reputation of scientists is evaluated.

Greene's television shows brought string theory to life, making abstract and speculative ideas vivid and real. These shows and his writings and his high media profile helped legitimate string theory in the public eye. This presentation of string theory as a glamorous field, its detractors argue, attracted talented students and research funding—enhancing its authority and status within physics. In subtle ways, factors outside string theory influenced to some degree how the field operated.

Gaia is inseparable from its creator, and Lovelock's fame helped it exert, in the end, a significant influence on scientific thought. Frustrated with the repeated failure to get his peers interested in the idea, Lovelock took it to *New Scientist*, where it sparked the attention of publishers, and *Gaia* not only outlined his scientific theory, but also provided parts of the public with a religious, spiritual, New Age, and ecological manifesto. This demand and interest from the public kept the idea alive when, Lovelock said, it was impossible to publish a paper in a scientific journal about Gaia.

But as scientific and political concerns over climate change grew, Gaia provided an example of a large-scale, systemic way of thinking about something as vast, something with so many moving parts, as global climate. Gaia provided a metaphor, too, for thinking about how the Earth would respond to threats, as it were, against "itself." The scientific establishment absorbed these ideas into a new discipline. As *Science* wrote in 2001, the threat of climate change meant it was "hard to imagine a more important discipline than Earth System Science."[8]

Gaia had as its subtitle *A New Look at Life on Earth*. And the scientific value of popular science is that it can provide a new way of viewing facts and concepts. This was Dawkins's aim with *The Selfish Gene* and, since his hands-on work in his early career, it has been his signature way of doing science. The method influenced scientists, causing a revolution in biology, as one scientist put it, because the clear articulation of the gene's-eye view forever altered the way biologists viewed evolution. Even though researchers were reluctant to cite it, a popular book, in formal research papers, a recent count on Google Scholar showed *The Selfish Gene* to have more than eighteen thousand citations. Staggering.

Gould's popular articulation of punctuated equilibrium gave the theory, its opponents asserted, a reputation among evolutionists that extended beyond its scientific merit. Pinker's writings on evolutionary psychology, journalists asserted, enhanced the status of the controversial field.

The writings of Pinker and Gould defy easy categorization. They wrote popular books as scholars at the pinnacle of their professions. Gould, similarly, in his books and columns often brought together the history of science and the social studies of science, and wanted his colleagues to cite his essays in their studies, as they would research papers published in scholarly journals. He was frustrated that other scholars did not do so, but yet, as one of his colleagues noted, this "hybridized style of nonpeer-reviewed but still scholarly publication . . . may be a viable genre for future 'scientific work.'"[9]

Pinker exemplified this point. He wrote that he sees little distinction between general and specialist readers, and so his books—all demanding, all readable, all evidence-based—can advance public understanding of science, but also the scientific understanding of new interdisciplinary areas like evolutionary psychology, cognitive science, and psycholinguistics. In *How the Mind Works*, he noted that this form of big-picture writing means that "there is little difference between a specialist and a thoughtful lay person because nowadays we specialists cannot be more than laypeople in most of our own disciplines, let alone neighboring ones."[10] In *Language, Cognition, and Human Nature* he wrote that he developed ideas in his popular books that could be examined in experiments. "These books," he wrote, "get more citations in the scholarly literature than my academic articles do."[11] *The Language Instinct*, for example, had by mid-2014 more than eight thousand five hundred citations on Google Scholar. *The Blank Slate* had more than three thousand. Both popular books exerted a deep influence on scientific and other scholarly communities.

Pinker also used his fame to introduce scientific knowledge into the public realm in novel ways. Rather than submitting a theoretical article to a journal, where it might wait a year to be published, he submitted it to a discussion forum at www.edge.org. The paper appeared the next day and, he recalled, compared to articles in a journal, was "subjected to far more rigorous scientific scrutiny."[12]

Celebrity has influenced science in other ways. When Tyson and the Hayden Planetarium left Pluto out of their public depiction of the solar system's planets, they catalyzed and influenced scientific discussion about how to define and categorize planets. The eventual decision of the community of astronomers agreed with Tyson's classification.

When Hawking became a scientific star, physicist Jeremy Dunning-Davies argued that colleagues of his had papers rejected by journals "simply because the end result disagrees with Hawking." He wrote: "papers which challenge Hawking in purely scientific grounds are not successful because his reputation has in some sense gone beyond the purely scientific." Hawking got to make a last-minute appearance at a major physics conference to discuss the ongoing puzzle that is the information paradox chiefly because he was, well, Stephen Hawking.

Crossover academic stardom enhances a scientist's value. Pinker became such a valued academic commodity that he was poached from MIT to Harvard. Pinker is a star attraction at fundraising events. At Hawking's seventieth birthday celebration, the head of Cambridge University twice asked for donations.

Stardom is seeping into all aspects of science. And that benefits science. Stardom can bring the values and interests of the public into science. The debate stirred up by these figures shows science that it does not stand on a pedestal and that it not somehow cut off from public debate and discussion.

A NEW SOURCE OF
SCIENTIFIC AUTHORITY: CELEBRITY

These scientists became celebrities with the power to influence citizen understanding, public culture, *and* scientific life. As a result, their scientific fame has developed into a unique, modern source of authority.

Their fame gave them a unique authority to discuss big ideas. In the process, they addressed nonscientific topics, such as religion, philosophy, history, sociology, culture, and childrearing. Unsurprisingly, this has led to criticism that the star scientists are suffering from a case of so-called expertise-creep, where their profile gives them an unwarranted ability to weigh in on issues about which they have little or no special expertise—and yet expect their opinions to matter by virtue of their reputations as "smart people."

Hawking said philosophy was dead and God was not needed to explain the universe—and he was savaged by philosophers, theologians, and, interestingly, by other scientists. Dawkins was dismissed as an amateur theologian. Pinker's writing on culture was criticized by humanists. Greenfield has been criticized for extending her expertise beyond neurodegenerative diseases to neuropsychology. Despite his output in the field, Gould felt he was never rated by historians of science.

But this type of critique is a predictable objection to public intellectual work, a gripe so entrenched that it is an unavoidable reaction to someone who speaks about general social and cultural concerns. The criticism ignores the landscape of modern science where ideas pollinate across fields and disciplines, creating new types of scientific knowledge. And it ignores a dimension of modern knowledge where ideas from the sciences can be applied to disciplines traditionally viewed as the domain of humanities. Theologians can study the nature of faith, sociologists can study the social influence of religion, and scientists can examine the psychology of religious belief.

In this intellectual climate, the scientists who become public intellectuals are brilliant synthesizers of science. With their established reputations, they can step back and provide a big-picture description of science, and can go on to explain it and describe its impact on our lives. Their celebrity helps give them a public platform to discuss these larger issues.[13] Major scientific celebrities have the legitimacy to discuss big ideas.

On the back of his popular scientific books, Pinker, for example, took his exploration of human nature into new fields. *The Better Angels of Our Nature* spanned a range of fields and was discussed by historians and political philosophers. It is not likely (I contest) that another cognitive scientist early in their career would be able to publish a similar book. Pinker had the track record, the following, and an established reputation as a brilliant interdisciplinary researcher to pull it off.

The evidence, likewise, in Gould's *The Mismeasure of Man* was based on his unique ability to combine the skills of the scientist and the historian, to reconstruct experiments and the context in which those experiments were performed. It is a powerful way to create new understandings and demands a significant level of expertise in science and beyond science. Only he could have written it.

The fame of the star scientists allowed them to introduce ideas into the public arena that have not yet been validated by scientific peers. Greene said that introducing this type of science-in-the-making is vital for enhancing science literacy. Moreover, this is often the type of knowledge that most addresses public concerns, where personal and political decisions must be made without knowledge that is 100 percent certain.

An intriguing example is Susan Greenfield's work that framed digital media as a social problem. Her title of baroness allowed her to raise issues about the links between digital media and conditions such as autism in Parliament, in newspapers, and in popular books. Yet detractors ridiculed her ideas because they lacked evidence—evidence she said would only appear in a convincing way in twenty years' time—and because she had not published on the topic in scientific venues. Yet she is an expert in the neurology of the brain who can read and synthesize neuroscience literature. For detractors, she either abuses her status to scaremonger, self-promote, or undermine the scientific method. For supporters, she uses her public position to hypothesize in a way that other scientists cannot.

Her interventions are valuable, however, because they bring into public discussion ideas of what constitutes valid evidence; what is meant by causality and correlation; and, most importantly, what advice science can give about behavior and policy in a situation of high uncertainty and no scientific consensus. At the same time, the public looks to science for answers or insights—and therefore scientists cannot be expected to stay behind the walls of their disciplines. A contribution that the modern celebrity scientist can make is to speak out on these issues, while public concern is high and all the evidence is still not in.

Dawkins also showed how scientific fame can spark social movements. *The God Delusion* was a foundational text in new atheism, which gathered a global community of atheists around a set of culturally significant books. Dawkins became the movement's driver and figurehead, providing an online space where like-minded atheists—deprived of an outlet, network, or prominent voice elsewhere—could meet. Dawkins's media work sparked the movement; his online community maintained and grew it. His continued public advocacy, with the film *The Unbelievers* being the latest product, is a way to keep the movement in the public mind.

Star scientists take on multiple roles in society. All at the same time, Hawking publishes influential papers; draws attention to political issues, such as his boycott of an Israeli conference in sympathy with Palestine; writes books explaining the universe that become instant best sellers; draws attention to a new research collaboration on tracheotomies; and garners tons of media attention to Irish bookmaker Paddy Power, as he created the formula that determined England's chances of success in the World Cup. Another physicist called the formula "meaningless" and "utterly pointless."[14]

Tyson, similarly, testified before Congress, helped write about teaching evolution in schools, advocated for increased funding for the space program, and drew eight million viewers to *Cosmos*. He appears on late-night talk shows, tweets to more than two million followers, writes books, and is interviewed in publications from the *New York Times* to *Parade* magazine. He advocates, advises, arbitrates, lobbies, clarifies, entertains—his fame allowing him to move between many roles. Doing so, he illustrates a major way that star scientists have influence: they reach audiences in a fragmented media environment. Celebrity draws a crowd.

Throughout the twentieth century, scientists have communicated, influenced policy, offered advice, and driven controversies. But what is different today is that celebrity grants scientists the authority to simultaneously undertake a variety of public roles without suffering a loss of scientific authority. The Sagan Effect—which claimed that successful popularizers were written off as poor scientists—has almost vanished. The celebrity scientists use their fame as a passport to access different influential arenas. Their celebrity is now its own distinct *source of authority*.

The last two scientists profiled here demonstrate this clearly. Greene moves seamlessly between science and entertainment, his fame granting him respect in both worlds. Tyson's stardom affects not only public understanding, but debates in his field and science policy. His scientific celebrity allows him to move between these different arenas of public life.

THE EMBLEMS OF A NEW SCIENTIFIC ERA

The scientists profiled in these pages show that the scientists who successfully communicate over a long period of time with broad audiences will inevitably become celebrities. They will likely follow the path of the public intellectual, then the mechanisms of celebrity will start to whir, and they will become celebrity intellectuals. As star scientists, they share the common characteristics of the scientists described here. Their media image will blur their public and private selves, they will become cultural commodities, they will embody reason and science, and they will receive some criticisms of their public role, especially as they are likely to speak outside their expertise.

That will carry risks for scientists. Hawking and Greenfield, especially, show the line between some private revelation and full public disclosure is quickly crossed. The loss of a completely private life has been the tax on their fame. And Greenfield demonstrates how female scientists still struggle in the straitjacket of historical portrayals of female scientists.

Yet Pinker, for example, shows that a scientist can maintain privacy while satisfying public demand for insights into the person behind the ideas. We know he likes biking and photography. We glimpsed inside his lovely Boston apartment. We can look into his genome, but don't learn as much as genetic evangelists would lead us to believe. What we really know about him can be found in his books. Gould managed this boundary also, with the two most intimate revelations—about his cancer and his son's autism—revealed to illustrate scientific ideas.

The Sagan Effect survives as a rhetorical stick to beat public scientists. Greenfield is less a scientist and more a marketing executive. Dawkins is an armchair scientist. Hawking is a great embarrassment to the scientific establishment. Asked about whether the Sagan Effect applied to his own career, Steven Pinker said his papers and grants were rejected before he became a success in popular culture, and his grants and papers are still rejected now . . . so it's impossible to tell.[15]

A clear indication of the impact of this fame culture is that science is now throwing up its tailor-made stars. Brian Cox, professor of particle physics at the University of Manchester, is for one newspaper "the new, young . . . face of science in Britain today." He took an original path to scientific fame. As his agent described, he "began his career not as a physicist but as a rock star."[16] In the late 1980s he signed a record deal with the band Dare, released two albums, toured with Jimmy Page and Europe. He then joined the band D:Ream and their song "Things Can Only Get Better" became the 1997 election anthem for Tony Blair. During these years, Cox earned a degree in physics and a doctorate in high-energy particle physics.

A strong scientist, he also had a talent for communication. From the mid 2000s, he presented a series of episodes of the BBC science series *Horizon*. He was science adviser to the 2007 film *Sunshine* where astronaut-scientists travel to reignite a dying sun. In 2010 he presented "Wonders of the Solar System," followed by "Wonders of the Universe" and "Wonders of Life." He is perfect for television, "a vision of gleaming skin, artfully floppy hair and extremely good teeth," wrote the *Daily Mail*, adding that women found him so attractive that he was called "Prof. Cox the Fox."[17] The number of applications to study science at college soared as students were inspired by his work. Commentators called this "The Cox Effect."[18]

He also conforms to a pattern where celebrity scientists often tend to be physicists, showing that despite the atomic destruction of World War II, physics still remains associated in the public mind with pure truth and physicists remain the custodians of that precious knowledge. Cox is a credentialed scientist, a clear communicator, and a researcher at CERN, the world's largest particle physics laboratory and the most impressive cathedral of modern science. But unlike Hawking or Greene he does not embody a new type of physics. He is, rather, a telegenic salesman for wonder, a machine-tooled mouthpiece for the glory of science. He is the first star scientist to emerge from within a developed celebrity culture—a minor music star who became a major celebrity scientist.

A criticism is that the rise of the celebrity scientists has impoverished public life. A select group of media-savvy scientists, the argument runs, has come to exert a cartel-like hold on science communication. Their stardom has also led to a cultural environment where discussions about the implications of science in society are reduced to a contest between the voices of established personalities, excluding other voices from public life.[19]

But this has not happened. Their cultural works and media coverage, wrapped around their public personalities, catalyzed wider discussions of science. Readers, as a result, are brought into a wider cultural conversation about the impact of scientific discoveries, the justification for scientific work, and the value of science in society.

The celebrity scientists are emblems of a new era of science, one embedded in the dynamics of the media, the needs of celebrity culture, and the vicissitudes of public life. They are powerful figures, who influence the public understanding of science *and* the future direction of society *and* the inner workings of science. They are science's new influential leaders, the ultramodern scientific elite—the new celebrity scientists.

Notes

CHAPTER ONE

1. Janet Browne, *Charles Darwin: The Power of Place* (London: Pimlico, 2003), 335. On Darwin's fame, see also: Janet Browne, "Charles Darwin as a Celebrity," *Science in Context* 16, nos. 1–2 (2003): 175–94, and Janet Browne, "Darwin in Caricature: A Study in the Popularisation and Dissemination of Evolution," *Proceedings of the American Philosophy Society* 145, no. 4 (2001): 496–509. When it was first published in 1859, Darwin's book was called *On the Origin of Species*. A sixth edition published in 1872 changed the title to *The Origin of Species*. The 150th anniversary edition was also called *The Origin of Species*. I use that title here because it is more readable.

2. On Einstein's fame, see John D. Barrow, "Einstein as Icon: How Einstein Became the Personification of Physics," *Nature* 433 (2005): 218–19; Marshall Missner, "Why Einstein Became Famous in America," *Social Studies of Science* 15, no. 2 (1985): 267–91; Roger Highfield and Paul Carter, *The Private Lives of Albert Einstein* (New York: St. Martin's Press, 1993); Abraham Pais, *Albert Einstein Lived Here* (Oxford: Oxford University Press, 1994); Alan J. Friedman and Carol C. Donley; *Einstein as Myth and Muse* (Cambridge University Press, 1985), 5–14, 18.

3. József Illy, ed., *Albert Meets America: How Journalists Treated Genius during Einstein's 1921 Travels* (Baltimore: Johns Hopkins University Press, 2006).

4. Missner, "Why Einstein," 268.

5. Pais, *Albert Einstein Lived Here*, 138.

6. Jane Gregory, *Fred Hoyle's Universe* (Oxford: Oxford University Press, 2005), 61.

7. Jon Agar, "What Happened in the Sixties?" *British Journal for the History of Science* 41 (2008): 567–600.

8. Declan Fahy and Bruce V. Lewenstein, "Scientists in Popular Culture: The Making of Celebrity," in *Routledge Handbook of Public Communication of Science and Technology*, ed. Brian Trench and Massimiano Bucchi (New York: Routledge, 2014).

9. Frederic Golden, "Showman of Science," *Time*, October 1980, 69.

10. On Sagan's science, career, and celebrity, see Keay Davidson, *Carl Sagan: A Life* (New York: John Wiley & Sons, 1999), and William Poundstone, *Carl Sagan: A Life in the Cosmos* (New York: Owl Books, 1999).

11. Michael Shermer, "This View of Science: Stephen Jay Gould as Historian of Science and Scientific Historian, Popular Scientist and Scientific Popularizer," *Social Studies of Science* 32 (2002): 489–524.

12. Rae Goodell, *The Visible Scientists* (Boston: Little Brown, 1977).

13. Robert K. Merton, "The Matthew Effect in Science," *Science* 159 (1968): 56–63.

14. Leo Braudy, *The Frenzy of Renown: Fame and Its History*, 2nd ed. (Oxford: Oxford University Press, 1997), 604.

15. Braudy, *The Frenzy of Renown*, 16.

16. Francesco Alberoni, "The Powerless 'Elite': Theory and Sociological Research on the Phenomenon of the Stars," in *Sociology of Mass Communications*, ed. Denis McQuail (London: Penguin, 1972).

17. David Foster Wallace, *Consider the Lobster and Other Essays* (London: Abacus, 2005), 143.

18. On celebrity as a potent and pervasive cultural phenomenon, see Graeme Turner, *Understanding Celebrity* (Sage: London, 2004), and P. David Marshall, *Celebrity and Power: Fame in Contemporary Culture* (Minneapolis: University of Minnesota Press, 1997).

19. "Science Guru Captured Popular Imagination," *Australian*, December 30, 1996, 10.

20. Cited in Davidson, *Carl Sagan*, 330.

21. This three-part definition of celebrity is taken from Turner, *Understanding Celebrity*.

22. Bill Hendrick, "Pop Scientist," *Atlanta Journal and Constitution*, December 21, 1996, 01F.

23. Louis Menand, "The Iron Law of Stardom," *New Yorker*, March 7, 1997, 36–39.

24. Jessica Evans and David Hesmondhalgh, *Understanding Media: Inside Celebrity* (Maidenhead, UK: Open University Press, 2005).

25. Rachel Carson, *Silent Spring*, 7th ed. (London: Penguin, 2000).

26. Keay Davidson, "Why Science Writers Should Forget Carl Sagan and Read Thomas Kuhn: On the Troubled Conscience of a Journalist," in *The Historiography of Contemporary Science, Technology, and Medicine*, ed. Ronald E. Doel and Thomas Söderqvist (London, Routledge, 2006).

27. John Ziman, *Real Science: What It Is, and What It Means* (Cambridge: Cambridge University Press, 2000); George Sarton, 1956, cited in Denis R. Alexander and Ronald L. Numbers (eds.), *Biology and Ideology from Descartes to Dawkins* (Chicago: University of Chicago Press, 2010), 1.

28. Thomas Kuhn, *The Structure of Scientific Revolutions*, 3rd ed. (London: University of Chicago Press, 1996).

29. Chargaff 1968 cited in Agar, "What Happened in the Sixties?" 590. For recent trends in science and its public presentation, see: Massimiano Bucchi, "Norms, Competition and Visibility in Contemporary Science: The Legacy of Robert K. Merton," *Journal of Classical Sociology* (2014), DOI: 10.1177/1468795X14558766.

30. Jean-François Lyotard, *The Postmodern Condition: A Report on Knowledge*, translated by G. Bennington and B. Massumi (Minneapolis: University of Minnesota Press, 1984).

31. Gerald Holton, *Einstein, History, and Other Passions: The Rebellion against Science at the End of the Twentieth Century* (New York: Basic Books, 1996).

32. Jon D. Miller, "Civic Scientific Literacy: The Role of the Media in the Electronic Era," in *Science and the Media*, eds. Donald Kennedy and Geneva Overholser (Cambridge, MA: American Academy of Arts and Sciences, 2010).

33. Matthew Nisbet and Ezra M. Markowitz, "Understanding Public Opinion in Debates over Biomedical Research: Looking Beyond Political Partisanship to Focus on Beliefs about Science and Society," *PLOS ONE*, 9, no. 2 (2014); Dietram A. Scheufele, "Communicating Science in Social Settings," *Proceedings of the National Academy of Sciences* 110 Supplement 3 (2013): 14040–47.

34. House of Lords Select Committee on Science and Technology, *Science and Society*, 3rd Report (London: HMSO, 2000).

35. Jane Gregory and Steve Miller, *Science in Public: Communication, Culture, and Credibility* (New York: Plenum, 1998), 1.

36. John Durant, "What Is Scientific Literacy?" *European Review* 2, no. 1 (1994): 83–89, 87. For an overview of recent global trends in scientific literacy, see: Martin W. Bauer, Rajesh K. Shukla, and Nick Allum, eds., *The Culture of Science: How the Public Relates to Science across the Globe* (New York: Routledge, 2012).

37. Bruce V. Lewenstein, "How Science Books Drive Public Discussion," in *Communicating the Future: Best Practices for Communication of Science and Technology to the Public*, ed. Gail Porter (Gaithersburg, MD: National Institute of Standards and Technology, 2002). Danette Paul, "Spreading Chaos: The Role of Popularizations in the Diffusion of Scientific Ideas," *Written Communication* 21, no. 1 (2004).

38. For a discussion on the ways science and the media are becoming closer, see Simone Rödder, Martina Franzen, and Peter Weingart, eds., *The Sciences' Media Connection: Public Communication and Its Repercussions* (New York: Springer, 2012); and David A. Kirby, *Lab Coats in Hollywood: Science, Scientists, and Cinema* (Cambridge, MA: MIT Press, 2011).

39. Stefan Collini, *Absent Minds: Intellectuals in Britain* (Oxford: Oxford University Press, 2006), 52.

40. The terms *professional culture* and *public culture* are described in detail by historian Thomas Bender in his *Intellect and Public Life: Essays on the Social History of Academic Intellectuals in the United States* (Baltimore: Johns Hopkins University Press, 1993).

41. Edward W. Said, *Representations of the Intellectual* (New York: Pantheon Books, 1994), 11.

42. Said, *Representations*, 12.

43. Said, *Representations*, 12.

44. Bryan Appleyard, "Two Cultures: A Science Fiction," *Independent*, September 27, 1995.

45. Bernard Dixon, "Sexy Science," *Current Biology* 7 (1997): 396.

46. Amanda Schaffer, "Science as Metaphor," *Slate*, July 6, 2004.

47. Stephen Jay Gould, *Full House: The Spread of Excellence from Plato to Darwin* (New York: Three Rivers Press, 1996), 8. This is a critical realist position, and for more on this idea, see Roy Bhaskar, *A Realist Theory of Science* (London: Verso, 2008). Sarah Tinker Perrault in *Communicating Popular Science: From Deficit to Democracy* (New York: Palgrave Macmillan, 2013) called this perspective a realist-skeptical viewpoint, and said it was a good one for science writers to hold. I agree.

48. I chose these four types of media by drawing on the influential work of film studies scholar Richard Dyer. In his book *Stars*, he argued the public persona of the film star is created over time as four categories of writings and visual images merge. First are films, which present the most important image of the actor. Second is promotion, where public relations professionals create a public image for the actor. Third is publicity, which consists of media profiles, interviews, and news reports. Last comes criticism and commentary, consisting of reviews, biographies, essays, and extended profiles in collected books, as well as fictional portrayals. For this book, I adapt Dyer's framework, replacing the category of films with that of the scientists' original writings. For more see: Richard Dyer, *Stars* (London: BFI Publishing, 1998). I requested an interview from all living scientists I profiled. I interviewed Steven Pinker and Susan Greenfield. The other scientists declined or did not respond to my request.

49. For details on "The Science Hall of Fame," see http://fame.gonzolabs.org/home.

CHAPTER TWO

1. Alok Jha, "Hawking at 70," *Guardian*, January 9, 2012, 3.

2. See Jenny Turner, "Scientific Sex Appeal," *Vogue*, April 1997, 42; Elizabeth Leane, *Reading Popular Physics: Disciplinary Skirmishes and Textual Strategies* (Aldershot, UK: Ashgate, 2007), 132; Arthur Lubow, "Heart and Mind," *Vanity Fair*, June 1992, 74.

3. On the history of cosmology, see Helge Kragh, *Quantum Generations: A History of Physics in the Twentieth Century* (Princeton, NJ: Princeton University Press, 2002).

4. William H. Cropper, *Great Physicists: The Life and Times of Leading Physicists from Galileo to Hawking* (New York: Oxford University Press, 2001), 456.

5. "Science: Those Baffling Black Holes," *Time*, September 4, 1978.

6. On this process, see Massimiano Bucchi, *Science and the Media: Alternative Routes to Scientific Communications* (New York: Routledge, 2012).

7. "Did a Black Hole Collide with the Earth in 1908," *Science News*, September 22, 1973, 180.

8. Stephen W. Hawking, "The Quantum Mechanics of Black Holes," *Scientific American*, January 1977, 34–40.

9. Hawking, "The Quantum Mechanics of Black Holes," 34.

10. David Abrahamson, "Magazine Exceptionalism: The Concept, the Criteria, the Challenge," *Journalism Studies* 8, no. 4 (2007): 667–70.

11. John Boslough, "The Unfettered Mind," *Science* 81, no. 2 (1981): 71.

12. Ian Ridpath, "Black Hole Explorer," *New Scientist*, May 4, 1978, 307.

13. Michael Harwood, "The Universe and Dr. Hawking," *New York Times Magazine*, January 23, 1983, 16–64.

14. Timothy Ferris, "Mind over Matter," *Vanity Fair*, June 1984, 56.

15. Dennis Overbye, "The Wizard of Space and Time," *Omni*, February 1979, 45.

16. Ridpath, "Black Hole Explorer," 309.

17. "Science: Soaring across Space and Time, *Time*, September 4, 1978.

18. Boslough, "The Unfettered Mind," 73.

19. Overbye, "The Wizard of Space and Time," 45.

20. Boslough, "The Unfettered Mind," 66.

21. Ferris, "Mind over Matter," 58.

22. Timothy Ferris, "Earth and Air, Fire and Water," *New York Times*, December 2, 1984, 76.

23. Rob Iliffe, "Isaac Newton: Lucatello Professor of Mathematics," in *Science Incarnate: Historical Embodiments of Natural Knowledge*, ed. Christopher Lawrence and Steven Shapin (Chicago: University of Chicago Press, 1998), 123.

24. Christopher Lawrence and Steven Shapin, "Introduction," in *Science Incarnate: Historical Embodiments of Natural Knowledge*, ed. Christopher Lawrence and Steven Shapin (Chicago: University of Chicago Press, 1998), 10, emphasis in original.

25. Kragh also noted that the end-of-physics theme has recurred in the history of the field. She noted that physics can end in a way that has nothing to do with final theories, but can finish when "people lose interest in it or because governments and funding agencies decide that there is no need to support basic physics research on a substantial scale." Kragh, *Quantum Generations*, 408.

26. Stephen Hawking, *Black Holes and Baby Universes and Other Essays* (London: Bantam Books, 1994), 30.

27. Cited in Judy Bachrach, "A Beautiful Mind, an Ugly Possibility, *Vanity Fair*, June 2004, accessed November 20, 2008, http://www.vanityfair.com/fame/features/2004/06/hawking200406.

28. Albert N. Greco, *The Book Publishing Industry* (Mahwah, NJ: Lawrence Erlbaum, 2005), 81.

29. Bernard Ryan, *Stephen Hawking: Physicist and Educator* (New York: Ferguson, 2005).

30. Stephen Hawking, *A Brief History of Time: From the Big Bang to Black Holes* (London: Bantam, 1988), vi.

31. Hawking, *A Brief History*, 175.

32. Hawking, *Black Holes*, 33.

33. Stephen Hawking, *A Brief History of Time: From the Big Bang to Black Holes* (New York: Bantam, 1988).

34. Stephen Hawking, *My Brief History* (New York: Bantam Books, 2013); Michael Rodgers, "The Hawking Phenomenon," *Public Understanding of Science* 1 (1992): 231–34.

35. Martin Gardner, The Ultimate Turtle," *New York Review of Books*, June 16, 1988.

36. Robyn Williams, "Genius Unique, Tragic and Triumphant," *Sydney Morning Herald*, July 2, 1988, 74.

37. Marcia Bartusiak, "What Place for a Creator?" *New York Times*, April 3, 1988, 10.

38. Hawking, *Black Holes*, 33.

39. Jon Turney, "The Word and the World: Engaging with Science in Print," in *Communicating Science: Contexts and Channels*, ed. Eileen Scanlon, Elizabeth Whitelegg, and Simeon Yates (London: Open University/Routledge, 1999).

40. Charles Oulton, "Cosmic Writer Shames Book World," *Sunday Times*, August 28, 1988.

41. Gilbert Adair, "Tale of the Unexpected," *Guardian*, November 5, 1997, 14.

42. Leane, *Reading Popular Physics*, 133.

43. Patricia Fara, *Newton: The Making of Genius* (London: Picador, 2002), 269.

44. Cited in Rodgers, "The Hawking Phenomenon," 233.

45. Kitty Ferguson, *Stephen Hawking: Quest for a Theory of Everything* (London: Bantam Books, 2001).

46. Michael White and John Gribbin, *Stephen Hawking: A Life in Science* (London: Abacus, 2003).

47. Anthony Burgess, "Towards a Theory of Everything," *Observer*, December 29, 1991, 42.

48. Bernard Carr, "Brief Histories of Hawking," *Independent*, January 5, 1992, 24.

49. Dennis Overbye, *Lonely Hearts of the Cosmos: The Quest for the Secret of the Universe* (London: Picador, 1993), 116.

50. Overbye, *Lonely Hearts*, 118.

51. *A Brief History of Time*, directed by Errol Morris (UK: Anglia Television Ltd/Gordon Freedman Productions, 1992), VHS.

52. Ryan, *Stephen Hawking*, 74.

53. Lubow, "Heart and Mind," 74.

54. Lubow, "Heart and Mind," 76–86.

55. Hawking, *Black Holes*, 2–20.

56. "Best Sellers: 7 November 1993," *New York Times*, November 7, 1993, 30.

57. "Beam Me Up, Stephen," *Evening Standard*, April 2, 1993, 8.

58. Hawking, *Black Holes*, 34.

59. "Emotional Turmoil of a Genius," *Toronto Star*, September 24, 1995, E2.

60. Jane Hawking, *Music to Move the Stars: A Life with Stephen* (London: Pan Books, 2000), 218, 298–99.

61. Helen Elliott, "Memoirs of a Blushing Bride," *Sydney Morning Herald*, January 15, 2000, 12.

62. Magdalen Ng, "My Life with a Genius," *Straits Times*, February 20, 2011.

63. Tim Adams, "Brief History of a First Wife," *Observer*, April 4, 2004, 4.

64. Elizabeth Grice, "A Brief History of Marriage," *Daily Telegraph*, October 20, 2006, 23.

65. Jeremy Dunning-Davies, "Popular Status and Scientific Influence: Another Angle on 'The Hawking Phenomenon,'" *Public Understanding of Science* 2 (1993): 85.

66. Dunning-Davies, "Popular Status," 86.

67. "Physics: Past, Present, Future," *Physics World*, December 1999, accessed March 10, 2010, http://physicsworld.com/cws/article/print/851. Einstein was top with 119 votes, followed by Newton with 96. Maxwell (67), Bohr (47), Heisenberg (30), Galileo (27), Feynman (23), Dirac (22), and Schrödinger (22) all appeared in the top ten.

68. Cited in "Physics: Past, Present, Future."

69. Robert Matthews, "Who's Afraid of Stephen Hawking?" *Spectator*, June 10, 2004.

70. Peter Coles, *Hawking and the Mind of God* (Cambridge: Icon Books, 2000), 61.

71. Jim Schembri, "Stephen Hawking's Universe," *The Age*, February 19, 1998, 2.

72. Walter Goodman, "Humankind's 3,000 Years of Prowling the Universe," *New York Times*, October 13, 1997, E6.

73. John Horgan, *The End of Science: Facing the Limits of Knowledge in the Twilight of the Scientific Age* (Reading, MA: Addison-Wesley, 1996), 94.

74. Jon Turney, "Strung Out," *Guardian*, November 10, 2001, 8.

75. John Gribbin, "Theories of Nearly Everything," *Independent*, November 3, 2001, 11.

76. Charles Arthur, "The Crazy World of Stephen Hawking," *Independent*, October 12, 2001, 7.

77. Steve Connor, "Higgs v Hawking: A Battle of the Heavyweights That Has Shaken the World of Theoretical Physics," *Independent*, September 3, 2002, 3.

78. Neil deGrasse Tyson, *The Sky Is Not the Limit: Adventures of an Urban Astrophysicist* (Amherst, NY: Prometheus Books, 2004).

79. Mark Henderson, "Scientific Minds Collide in Matter of Hawking's Fame," *Times* (London), September 4, 2002, 12.

80. Owen Gingerich, "Looking Up to the Stars," *Nature* 421 (2003): 694.

81. Marion Frith, "Just What Is Going on in the Hawking Home?" *Sunday Age*, February 1, 2004, 10.

82. Paul Sims, "Hawking Daughter: How Claims of Abuse Drove Me to Drink," *Evening Standard*, April 13, 2004, 20.

83. Steve Connor, "The IOS Profile: Stephen Hawking," *Independent on Sunday*, January 25, 2004, 23.

84. Andy Dolan, "Mrs Hawking Says 'I Love You' with a Red Balloon," *Daily Mail*, February 16, 2004, 9.

85. Ian Burrell, "Hawking Rejected BBC Drama Script as Soap Opera," *Independent*, December 5, 2003, 7.

86. Cited in Kate Holton, "Galactic Black Holes May Have Looser Grip; Stephen Hawking's Findings Shifts Earlier Work," *Philadelphia Inquirer*, July 16, 2004, A02.

87. Kitty Ferguson, *Stephen Hawking: An Unfettered Mind* (New York: Palgrave Macmillan, 2012).

88. Hicklin cited in Hélène Mialet, *Hawking Incorporated: Stephen Hawking and the Anthropology of the Knowing Subject* (Chicago: University of Chicago Press, 2012), 92.

89. Mialet, *Hawking Incorporated*, 116.

90. John A. Smolin and Jonathan Oppenheim, "Locking Information in Black Holes," *Physical Review Letters* 96, 081302-1.

91. Bryan Appleyard, "Give Hawking a Medal—But Not for His Cosmic Theory," *Sunday Times*, December 3, 2006, 6.

92. Mike Wade, "Higgs Launches Stinging Attack against Nobel Rival," *Times* (London), September 11, 2008, 8.

93. Rachel Cooke, "Master of His Universe," *Observer Magazine*, March 2, 2008, 14.

94. Michael Turner, "No Miracle in the Multiverse," *Nature* 467 (2010): 658.

95. Stephen Hawking and Leonard Mlodinow, *The Grand Design* (New York: Bantam Books, 2010), 5.

96. Adam Gabbatt, "Universe Not Created by God, Says Hawking," *Guardian*, September 2, 2010, 11.

97. Graham Farmelo, "Has Stephen Hawking Seen Off God?" *Telegraph*, September 3, 2010, 19.

98. Philip Ball, "The Hawking Delusion," *Prospect*, September 8, 2010, accessed July 16, 2014, http://www.prospectmagazine.co.uk/science-and-tech nology/the-hawking-delusion.

99. Cited in Hannah Devlin, "Hawking Has Got It Wrong, Says Colleague," *Times* (London), October 1, 2010, 26.

100. Steve Connor, "We Shouldn't Attach Any Weight to What Hawking Says about God," *Independent*, September 27, 2010, 14.

101. Craig Callender, "There Is No Theory of Everything," *New Scientist*, September 11, 2010, accessed July 16, 2014, http://www.newscientist.com/blogs/ culturelab/2010/09/stephen-hawking-says-theres-no-theory-of-everything.html.

102. Graham Farmelo, "Life, the Universe and M-theory," *Times* (London), September 11, 2010, 9.

103. Dwight Garner, "Many Kinds of Universes, and None Require God," *New York Times*, September 8, 2010, C1.

104. Cited in "Science Makes God Unnecessary: Stephen Hawking," *Korea Times*, September 8, 2010.

105. Turner, "No Miracle," 657.

106. John Crace, "The Grand Design," *Guardian*, September 14, 2010, 17.

107. Ian Sample, "There Is No Heaven or Afterlife . . . That Is a Fairy Story for People Afraid of the Dark," *Guardian*, May 16, 2011, 3.

108. Ferguson, *An Unfettered Mind*, 199.

109. Ferguson, *An Unfettered Mind*, 261.

110. Ferguson, *An Unfettered Mind*, 225.

111. Ferguson, *An Unfettered Mind*, 131. Also, the biography's opening chapter acknowledges, but does not resolve, apparent contradictions in his life and career. Ferguson wrote: "Hawking's life story and his science are rife with paradoxes. Things are not often what they seem." Ferguson, *An Unfettered Mind*, 4.

112. Ferguson, *An Unfettered Mind*, 178.

113. Ed Lake, "I Think; Therefore I'm Rich," *Daily Telegraph*, January 14, 2012, 29.

114. As an indicator of cultural profile, I set a classic Downsian pattern to media attention to Hawking, gathered from mentions of his name in the Lexis-Nexis database, searched year by year under major world publication headings. This was not a formal content analysis, but was designed to map in some way patterns of attention

to him over time. It should be noted that the archives from LexisNexis are often incomplete before 1980. (See Anthony Downs, "Up and Down with Ecology: The 'Issue-Attention Cycle,'" *Public Interest* 28 (1972): 38–50.

115. Gordon Rayner, "Stephen Hawking Says He Was 'Delighted' to Feature in Paralympics Opening Ceremony," Telegraph.co.uk, 91, August 30, 2012.

116. Hawking, *My Brief History*, 91.

117. Hawking, *My Brief History*, 121–22.

118. Mialet, *Hawking Incorporated*. The 2014 film *The Theory of Everything*, like the drama *Hawking*, presents a more rounded view of the physicist, giving audiences a glimpse of Hawking as a young man and husband. Not coincidentally, the film's source material is Jane Hawking's biographical descriptions of their life together.

CHAPTER THREE

1. Andrew Anthony, "God of the Godless," *Observer*, September 15, 2013, 18.

2. Richard Dawkins, *An Appetite for Wonder: The Making of a Scientist* (New York: Ecco, 2013).

3. Fern Elsdon-Baker, *The Selfish Genius: How Richard Dawkins Rewrote Darwin's Legacy* (London: Icon Books, 2009), 2.

4. Ginny Dougary, "Gene Genie," *Times* (London), August 31, 1996.

5. Jenny Diski, "Back to *The Naked Ape*," *Guardian*, October 15, 2011, 9.

6. Dawkins, *An Appetite for Wonder*, 275–81; Soraya de Chadarevian, "The Selfish Gene at 30: The Origin and Career of a Book and Its Title," *Notes and Records of the Royal Society* 61, no. 1 (2007): 31–38.

7. Cited in de Chadarevian, "*The Selfish Gene* at 30," 33.

8. John Pfeiffer, "*The Selfish Gene*," *New York Times*, February 27, 1977, 10.

9. Douglas R. Hofstadter, "The Romance of Science," *Washington Post*, December 2, 1979, 1.

10. Ullica Segerstråle, *Defenders of the Truth: The Sociobiology Debate* (Oxford: Oxford University Press, 2000).

11. Peter Gwynne, Sharon Begley, and Allan J. Mayer, "Our Selfish Genes," *Newsweek*, October 16, 1978, 118.

12. Steve Rose, Richard Lewontin, and Leon Kamin, *Not in Our Genes: Biology, Ideology and Human Nature* (London: Pantheon, 1984).

13. Oliver James, "Despite Ourselves, We Are All Gordon Gekkos Now," *Independent on Sunday*, January 27, 2008, accessed July 16, 2016, http://www .independent.co.uk/voices/commentators/oliver-james-despite-ourselves-we-are-all-gordon-gekkos-now-774535.html.

14. Mary Midgley, "Gene-Juggling," *Philosophy* 54, no. 210 (1979): 439–58.

15. Richard Dawkins, *The Selfish Gene* (Oxford: Oxford University Press, 1989), ix.

16. Alan Grafen and Mark Ridley (eds.), *Richard Dawkins: How a Scientist Changed the Way We Think* (Oxford: Oxford University Press, 2006), 72. The press's blurb for the thirtieth anniversary edition of the book said the "imaginative, powerful, and stylistically brilliant work not only brought the insights of Neo-Darwinism to a wide audience, but galvanized the biology community, generating much debate and stimulating whole new areas of research."

17. Richard Dawkins, *The Extended Phenotype: The Long Reach of the Gene* (Oxford University Press, 1999), vi.

18. John A. Endler, "Son of Selfish Gene," *Quarterly Review of Biology* 58 (1983): 224–27.

19. Richard E. Michod, "*The Extended Phenotype,*" *American Scientist* 71, no. 5 (1983): 526.

20. Christopher Wills, "Evolution by Metaphor," *Science* 218, no. 4577 (1982): 1109–10.

21. Hee-Joo Park, "The Creation-Evolution Debate," *Public Understanding of Science* 10 (2001): 173–86.

22. For a history of creationism, see Ronald L. Numbers, *The Creationists* (New York: Alfred A. Knopf, 1992); Chris Mooney, *The Republican War on Science* (New York: Basic Books, 2006); and Barry A. Palevitz, "Intelligent Design Creationism: None of Your Business? Think Again," *Evolution* 56, no. 8 (2002): 1718–20.

23. Richard Dawkins, *The Blind Watchmaker* (London: Penguin, 1991), xiv–xv, emphasis in original.

24. "*The Blind Watchmaker,*" *Sydney Morning Herald*, December 30, 1989.

25. Michael T. Ghiselin, "We Are All Contraptions," *New York Times*, December 14, 1986, 18.

26. Sarah Duncan, "The Zoology Man," *Times* (London), October 3, 1986.

27. Jon Turney, "Telling the Facts of Life: Cosmology and the Epic of Evolution," *Science as Culture* 10 (2001): 225–47.

28. Melvyn Bragg, "There Is Poetry in Science," *Observer*, October 18, 1998, 13; Daniel C. Dennett, *Darwin's Dangerous Idea: Evolution and the Meanings of Life* (New York: Simon & Schuster, 1995).

29. Dawkins would explore this theme further in *The Ancestor's Tale: A Pilgrimage to the Dawn of Evolution* (2004), his history of life on Earth that was for one writer "a kind of textbook in disguise." Jon Turney, "The Latest Boom in Popular Science Books," in *Journalism, Science and Society: Science Communication between News and Public Relations*, ed. Martin W. Bauer and Massimiano Bucchi (New York: Routledge, 2007), 89.

30. Andrew Brown, *The Darwin Wars: How Stupid Genes Became Selfish Gods* (London: Simon & Schuster, 1999), 21.

31. Robin McKie, "Survival of the Bitchiest as the Darwinian Bulldogs Go to War," *Observer*, September 27, 1998, 3.

32. Segerstråle, *Defenders of the Truth*, 324.

33. John Gribbin, "The Gene Genie," *Sunday Times*, May 21, 1995.

34. Tim Radford, "Astounding Stories," *Guardian*, July 17, 1996, T2.

35. Michael Ruse, "The Survival of the Evolutionists," *Globe and Mail*, December 12, 1998, D19.

36. John Horgan, *The End of Science: Facing the Limits of Knowledge in the Twilight of the Scientific Age* (London: Abacus, 1998), 116.

37. Richard Dawkins, *River out of Eden: A Darwinian View of Life* (London: Phoenix, 1995), 35–36.

38. Richard Dawkins, *A Devil's Chaplain: Selected Essays* (London: Phoenix, 2004), 22.

39. Dawkins, *A Devil's Chaplain*, 8.

40. Richard Dawkins, *Unweaving the Rainbow* (London: Penguin, 2006), 34–35.

41. Timothy Ferris, "Frauds! Fakes! Phonies!" *New York Times*, January 10, 1999, 7.

42. Colin Hughes, "Richard Dawkins: The Man Who Knows the Meaning of Life," *Guardian*, October 3, 1998, 6. The piece was also published as "The Evolution of Richard," *Sydney Morning Herald*, November 28, 1998, 35.

43. Paul Johnson, "Is This the Most Dangerous Man in Britain Today?" *Daily Mail*, February 1, 1999, 10.

44. Marek Kohn, *A Reason for Everything: Natural Selection and the English Imagination* (London: Faber & Faber, 2005), 319.

45. Simon Hattenstone, "Darwin's Child," *Guardian*, February 10, 2003.

46. Michael Shermer, "The Skeptic's Chaplain: Richard Dawkins as Fountainhead of Skepticism," in *Richard Dawkins: How a Scientist Changed the Way We Think*, ed. Alan Grafen and Mark Ridley (Oxford: Oxford University Press, 2006), 228.

47. Bryan Appleyard, "The Fault Is Not in Our Genes But in Our Minds," *Sunday Times*, November 28, 2004, 5.

48. Ruth Gledhill, "God . . . in Other Words," *Times* (London), May 10, 2007, 4.

49. Appleyard, "The Fault Is Not in Our Genes But in Our Minds," 5.

50. Robin McKie, "Doctor Zoo," *Observer*, July 25, 2004, 25.

51. Robert Macfarlane, "Articles of Faith," *Spectator*, February 15, 2003, accessed July 16, 2014, http://www.spectator.co.uk/books/20455/articles-of-faith/.

52. Brian Trench, "Towards an Analytic Framework of Science Communication Models," in *Communicating Science in Social Contexts: New Models, New Practices*, ed. Donghong Cheng, Michel Claessens, Toss Gascoigne, Jenni Metcalfe, Bernard Schiele, and Shunke Shi (Berlin: Springer, 2008), 121.

53. Dawkins, *Unweaving the Rainbow*, 22–23.

54. Eryn Brown, "Feeding an Insatiable 'Appetite for Wonder,'" *Los Angeles Times*, November 30, 2013, AA2.

55. Dawkins, *A Devil's Chaplain*, 43.

56. Richard Dawkins, "Religion's Misguided Missiles," *Guardian*, September 15, 2001, 20.

57. Decca Aitkenhead, "'People Say I'm Strident,'" *Guardian*, October 25, 2008, 31.

58. Thomas Nagel, "The Fear of Religion," *New Republic*, October 23, 2006.

59. H. Allen Orr, "A Mission to Convert," *New York Review of Books*, 54, January 11, 2007.

60. Terry Eagleton, "Lunging, Flailing, Mispunching," *London Review of Books*, October 19, 2006.

61. Daniel Dennett, "*The God Delusion*," *New York Review of Books*, 54, March 1, 2007.

62. Steven Weinberg, "A Deadly Certitude," *Times Literary Supplement*, January 19, 2007.

63. Ian McEwan, "A Parallel Tradition," *Guardian*, April 1, 2006.

64. Richard Dawkins, *The God Delusion* (London: Black Swan, 2007).

65. Robert C. Fuller, *Spiritual, but Not Religious: Understanding Unchurched America* (New York: Oxford University Press, 2001).

66. Richard Cimino and Christopher Smith, "The New Atheism and the Formation of the Imagined Secularist Community," *Journal of Media and Religion* 10 (2011): 24–38.

67. Michael Shermer, "Arguing for Atheism," *Science* 315, no. 5811 (2007): 463.

68. Trench, "Towards an Analytic Framework of Science Communication Models," 122.

69. Andrew Brown, "The New Atheism: A Definition and a Quiz," *Guardian*, December 29, 2008, accessed July 16, 2014, http://www.guardian.co.uk/comment isfree/andrewbrown/2008/dec/29/religion-new-atheism-defined. New atheism shares three core characteristics of social movements: they are in conflict with identifiable opponents, they are linked by informal networks, and they have a collective identity. See Donatella della Porta and Mario Diani, *Social Movements: An Introduction* (Malden, MA: Blackwell Publishing, 2006).

70. Cimino and Smith, "The New Atheism and the Formation of the Imagined Secularist Community," 37.

71. Gary Wolf, "The Church of the Non-Believers," *Wired*, November 2006, 182–93.

72. As an indicator of cultural profile, I set a classic Downsian pattern to media attention to Dawkins, gathered from mentions of his name in the Lexis-Nexis database, and searched year by year under major world publication headings. This

was not a formal content analysis, but was designed to map in some way patterns of attention to him over time. It should be noted that the archives from LexisNexis are often incomplete before 1980. See Anthony Downs, "Up and Down with Ecology: The 'Issue-Attention Cycle,'" *Public Interest* 28 (1972): 38–50.

73. Accessed July 3, 2009, http://www.youtube.com/user/richarddawkins dotnet.

74. Gordon Lynch, "Richard Dawkins, TV Evangelist," *Guardian*, August 11, 2007, accessed July 16, 2014, http://www.theguardian.com/commentisfree/2007/aug/11/atheismthenewzealotry.

75. George Johnson, "A Free-for-All on Science and Religion," *New York Times*, November 21, 2006, F1.

76. Agillesp123, *Dawkins vs. Tyson.* YouTube. Accessed October 8, 2012, http://www.youtube.com/watch?v=-_2xGIwQfik Dawkins Vs Tyson.

77. "Go God Go," *South Park*, first broadcast November 1, 2006.

78. Elsdon-Baker, *The Selfish Genius*, 2.

79. Richard Dawkins, "Extended Phenotype—but Not *Too* Extended: A Reply to Laland, Turner and Jablonka," *Biology and Philosophy* 19 (2004): 377–97.

80. European Science Foundation, "European Evolutionary Biologists Rally Behind Richard Dawkins's Extended Phenotype," January 20, 2009, http://www.sciencedaily.com/releases/2009/01/090119081333.htm.

81. Jerry A. Coyne, "His Tale Is True," *Times Literary Supplement*, June 16, 2006, 9; and Robert Aunger, "What's the Matter with Memes?" in *Richard Dawkins: How a Scientist Changed the Way We Think*, ed. Alan Grafen and Mark Ridley (Oxford: Oxford University Press, 2006).

82. James Gleick, *The Information: A History, a Theory, a Flood* (New York: Vintage, 2011), 312–16.

83. Daniel Trilling, "Beyond Dawkins," *New Humanist*, September/October 2013, accessed July 16, 2014, http://rationalist.org.uk/articles/4271/beyond-dawkins.

84. Nigel Farndale, "Has Richard Dawkins Found a Worthy Opponent at Last?" telegraph.co.uk, December 29, 2012.

85. Richard Dawkins, *The Greatest Show on Earth: The Evidence for Evolution* (London: Bantam, 2009), 437.

86. "Our Beginnings," Richard Dawkins Foundation, accessed July 16, 2014, http://richarddawkins.net/aboutus/.

87. Duncan, "The Zoology Man."

88. Richard Girling, "The Tracing of the Shrew," *Times* (London), August 13, 1994.

89. Matt Ridley, "A Biologist Whom People Love to Hate," *Times* (London), May 15, 1995.

90. Brenda Maddox, "Let's Say Goodbye to Frankenstein," *Times* (London), March 20, 1996.

91. Quentin Letts, "Faith, Hope and the Darwin Man," *Times* (London), March 29, 1996.

92. Andrew Billen, "Show Us the Monkey," *Times* (London), September 11, 2004, 8.

93. Alister McGrath, "The Enlightenment Is Over, and Atheism Has Lost Its Moral Cutting Edge," *Times*, October 29, 2005, 77.

94. Jonathan Sacks, "Danger Ahead—There Are Good Reasons Why God Created Atheists," *Times*, October 21, 2006, 80.

95. Matthew Parris, "The Terrorist Virus Is No Lightweight Matter," *Times* (London), November 11, 2006, 17.

96. Mark Henderson, "Science Vs. God: The Showdown," *Times* (London), November 25, 2006, 11.

97. Gledhill, "God . . . in Other Words," 4.

98. Andrew Billen, "Mr. Logic Takes on the Idiots," *Times* (London), August 14, 2007, 23.

99. Anjana Ahuja, "'Evolution Is God's Work,'" *Times* (London), June 4, 2009, 4.

100. Episode 5, *Genius of Britain: The Scientists Who Changed the World*, first broadcast June 3, 2010. (Dawkins's answer, by the way, was: "Well, I notice that you brought up the question of God and I didn't.")

101. Dawkins, *An Appetite for Wonder*, 166.

102. Brandon Robshaw, "Dawkins Gives an Insight Beyond the Caricature," *Independent*, September 15, 2013, 18.

103. Janet Maslin, "Science, Evidently, Was in His Genes," *New York Times*, September 19, 2013, C1.

104. Jenni Russell, "Charles Darwin and Me," *Sunday Times*, September 15, 2013, Culture, 40–41.

105. Dennis Overbye, "Intellectuals on a Mission," *New York Times*, December 10, 2013, D5.

106. Gary Goldstein, "'Unbelievers' Preaches to the Skeptical Choir," *Los Angeles Times*, November 27, 2013.

CHAPTER FOUR

1. Steven Pinker, "My Genome, My Self," *New York Times Magazine*, January 11, 2009, 30. Pinker's genome can be viewed at the PGP website at: https://my.pgp-hms.org/profile/hu04FD18, accessed October 22, 2014.

2. Dawkins's quote is from a review of Pinker's *The Blank Slate*, which appeared in 2002 in the *Times Literary Supplement* and was reproduced on the back cover of Pinker's *Language, Cognition, and Human Nature*. Steven Pinker,

Language, Cognition, and Human Nature: Selected Articles (New York: Oxford University Press, 2013).

3. For an accessible overview of Chomsky's ideas, see George Steiner, *George Steiner at "The New Yorker"* (New York: New Directions, 2009), 276–94.

4. See Michael Burleigh, *The Third Reich: A New History* (London: Pan Books, 2001).

5. Steven Pinker, *How the Mind Works* (New York: W. W. Norton, 1997), 426.

6. *Me & Isaac Newton*, directed by Michael Apted (USA: First Look Pictures, 2000), DVD.

7. Pinker, *Language, Cognition, and Human Nature*, 1.

8. Pinker was the sole author of *Learnability and Cognition: The Acquisition of Argument Structure* (1989), the editor of *Visual Cognition* (1985), and the coeditor of *Connections and Symbols* (1988) and *Lexical and Conceptual Semantics* (1992).

9. John Algeo, "Words and Rules," *Journal of English Linguistics* 28, no. 4 (2009): 593–95.

10. Pinker, *Language, Cognition, and Human Nature*, ix–x.

11. Steven Pinker, *The Language Instinct: How the Mind Creates Language* (New York: HarperPerennial Modern Classics, 2007).

12. Pinker, *The Language Instinct*, 5.

13. Pinker, *The Language Instinct*, 119. Pinker said he is often assumed to be a student or disciple of Chomsky. At Harvard he could take classes at MIT and he took a course taught by Chomsky on theories of the mind. Chomsky, Pinker later wrote in *Language, Cognition, and Human Nature* (228), was "a major intellectual influence" on him, but he distanced himself from some of Chomsky's theories of language and "the worshipful academic cult" that surrounded the older man. The most decisive intellectual influences on Pinker were Stephen Kosslyn, his primary doctoral mentor, and Roger Brown, the suave scholar Pinker later called "the Cary Grant of psychology," a social psychologist who made fundamental contributions to areas that Pinker would subsequently focus on, such as how children acquire language and how language relates to thought. See: Steven Pinker, "Obituary: Roger Brown," *Cognition* 66 (1998): 199.

14. Steven Pinker, "Survival of the Clearest," *Science* 404 (2000): 441–42.

15. Massimo Piattelli-Palmarini, "Speaking in Too Many Tongues," *Nature* 408 (2000): 403.

16. Thomas Wasow, "Words and Rules: The Ingredients of Language," *Language* 77, no. 1 (2001): 168.

17. Randy Harris, "The Popularization of Noam Chomsky," *Globe and Mail*, June 18, 1994.

18. Pinker, *The Language Instinct*, 8.

19. Howard Gardner, "Green Ideas Sleeping Furiously," *New York Review of Books*, March 23, 1995.

20. John Gribbin, "To Boldly Go," *Sunday Times* (London), April 10, 1994.

21. Pinker, *Language, Cognition, and Human Nature*, x.

22. David Mehegan, "Language's Bad Boys," *Boston Globe*, March 29, 1994, 69.

23. Steven Pinker, telephone interview with the author, March 4, 2014.

24. Mark Abley, "Language Expert, Montrealer's the Genuine Article," *Gazette*, April 2, 1994, A1.

25. Pinker, *The Language Instinct*, 406.

26. Anthony Gottlieb, "It Ain't Necessarily So," *New Yorker*, September 17, 2012.

27. Robert Wright, "The Evolution of Despair," *Time*, August 28, 1995.

28. Pinker later wrote that he is often erroneously identified as a researcher in evolutionary psychology. In fact, he said he published only one piece of original research within that framework. Pinker, *Language, Cognition, and Human Nature*.

29. Pinker, *How the Mind Works*, 52.

30. John Horgan, *The Undiscovered Mind: How the Human Brain Defies Replication, Medication, and Explanation* (New York: Touchstone, 2000).

31. Steven Johnson, "Sociobiology and You," *Nation*, November 18, 2002, 12–18.

32. Christopher Lehmann-Haupt, "Thinking Deeply About Thinking and Having Fun, Too," *New York Times*, November 24, 1997, E8.

33. Marek Kohn, "Undressing for Success," *Independent*, January 17, 1998, 10.

34. Oliver Morton, "Doing What Comes Naturally," *New Yorker*, November 3, 1997, 102.

35. Michael S. Gazzaniga, "How the Mind Works," *Trends in Cognitive Sciences* 2, no. 1 (1998): 38.

36. Steve Jones, "The Set within the Skull," *New York Review of Books*, November 6, 1997.

37. Anna Mundow, "Evolutionary Rock Star," *Irish Times*, January 3, 1998, 60.

38. Simon Garfield, "Mind Games with Kate, Karl and Groucho," *Mail on Sunday*, January 25, 1998, 40.

39. Elaine Showalter, "Something on His Mind," *Times* (London), January 15, 1998.

40. John Horgan, "Darwin on His Mind," *Lingua Franca* 7 (1997): 42.

41. Tim Cornwell, "A Stone Age Mind-Blower," *Times Higher Education Supplement*, January 9, 1998, 17.

42. Nigel Hawkes, "The Man Who Rewrote the Book on Language," *Times* (London), October 20, 1999.

43. Mark Aronoff, "Washington Sleeped Here," *New York Times*, November 28, 1999, 26.

44. George C. Williams, "Steven Pinker," in *The Third Culture: Beyond the Scientific Revolution*, ed. John Brockman (New York: Touchstone, 1995), 223.

45. David Poeppel, "Instincts for the Past Tense," *Nature* 403 (2000): 361.

46. Ed Douglas, "Steven Pinker: The Mind Reader," *Guardian*, November 6, 1999.

47. Steven Pinker, "Some Remarks on Becoming a 'Public Intellectual,'" MIT Communications Forum, accessed July 16, 2009, http://web.mit.edu/comm-forum/papers/pinker.html.

48. As an indicator of cultural profile, I mapped the pattern of media attention to Pinker gathered from mentions of his name in the LexisNexis database, searched year by year under major world publication headings. This was not a formal content analysis, but was designed to map in some way patterns of attention to him over time. It should be noted that the archives from LexisNexis are often incomplete before 1980. See Anthony Downs, "Up and Down with Ecology: The 'Issue-Attention Cycle,'" *Public Interest* 28 (1972): 38–50.

49. Steven Pinker, telephone interview with the author, March 4, 2014.

50. Steven Pinker, *The Blank Slate: The Modern Denial of Human Nature* (New York: Viking, 2002), vii–xiii.

51. Colin McGinn, "All in Our Heads," *Washington Post*, October 13, 2002, T3.

52. Pinker, *The Blank Slate*, ix.

53. Pinker, *The Blank Slate*.

54. John Morrish, "Why Scientists Think the Ghost Is Toast," *Independent on Sunday*, September 29, 2002, 17.

55. Robert J. Richard, "The Evolutionary War," *New York Times*, October 13, 2002, 9.

56. Johnson, "Sociobiology and You," 12–18.

57. Philip Gerrans, "Slaves to Our True Selves," *Sydney Morning Herald*, January 25, 2003, Books, 10.

58. Pinker, *The Blank Slate*, 384.

59. Robin McKie and Vanessa Thorpe, "Clash of the Titans," *Observer*, September 22, 2002, 18.

60. Nicholas Wade, "Scientist at Work: Steven Pinker," *New York Times*, September 17, 2002, F1.

61. Ben Schrank, "Mind Games," *Financial Times*, September 12, 2002, 3.

62. David L. Hull, "Nurturing a View of Human Nature," *Nature* 419 (2002): 252.

63. Simon Blackburn, "The Blank Slate," *New Republic*, November 25, 2002, 28.

64. Louis Menand, "Dangers Within and Without," *Profession*, December 2005, 10–17.

65. Robert Wright, "Steven Pinker," *Time*, April 26, 2004.

66. Patrick Healy, "Harvard Raids MIT for Eminent Professor," *Boston Globe*, April 4, 2003.

67. Karen Arenson, "Boldface Professors," *New York Times*, April 25, 2004, 22.

68. David Crystal, "We Are What We Say," *Financial Times*, October 6, 2007, accessed July 16, 2014, http://www.ft.com/intl/cms/s/0/1da850dc-73a7-11dc-abf0-0000779fd2ac.html#axzz37k13Q6k0.

69. David Papineau, "Caveman Conversations," *Independent on Sunday*, October 5, 2007, 22.

70. Deborah Cameron, "Talking Outside the Box," *Guardian*, October 6, 2007, 7.

71. William Saletan, "The Double Thinker," *New York Times*, September 23, 2007, 14.

72. Vivienne Parry, "The Man Who Swears by Popular Science," *Times* (London), October 20, 2007.

73. Bryan Appleyard, "Steven Pinker Knows What's Going On inside Your Head," *Sunday Times*, October 14, 2007.

74. Oliver Burkeman, "Basic Instincts," *Guardian*, September 22, 2007.

75. Pinker, *Language, Cognition, and Human Nature*, 302.

76. Steven Pinker, *The Better Angels of Our Nature: Why Violence Has Declined* (New York: Viking, 2011), xxi.

77. Pinker, *The Better Angels of Our Nature*, xxvii.

78. Martin Daly, "A Farewell to Arms," *Nature* 478 (2011): 454.

79. David Runciman, "Make Love, Not War," *Guardian*, September 24, 2011, 8.

80. In its original and narrowest sense, Whig history refers to political and religious history, in particular "an interpretation of British history, prevalent in Whig political and intellectual circles in the mid-nineteenth century, which stressed the growth of liberty, parliamentary rule and religious toleration since the constitutional struggles of the seventeenth century." Adrian Wilson and T. G. Ashplant, "Whig History and Present-centred History," *Historical Journal* 31, no. 1 (1988): 1–16.

81. Wesley Yang, "Nasty, Brutish, and Long," *New York*, October 24, 2011.

82. John Gray, "Delusions of Peace," *Prospect*, September 21, 2011.

83. Annie Maccoby Berglof, "At Home: Steven Pinker," *Financial Times*, December 14, 2012.

84. Joshua Kendall, "The Shutterbug Scientist," *Financial Times*, October 8, 2011, 2.

85. Steven Pinker, telephone interview with the author, March 4, 2014.

86. Steven Pinker, telephone interview with the author, March 4, 2014.

87. Michel Jean-Baptiste et al., "Quantitative Analysis of Culture Using Millions of Digitized Books," *Science* 331 (2011): 176–82.

88. Stephen Pinker, "Science Is Not Your Enemy," *New Republic*, August 6, 2013.

89. Another example is: "For almost ten years I was obsessed with solving a paradox in language acquisition. It began with a linguistic puzzle: Why do sentences such as *He poured water into the glass* and *He filled the glass with water* sound OK, but seemingly similar sentences such as *He poured the glass with water* and *He filled water into the glass* sound odd?" Pinker, *Language, Cognition, and Human Nature*, 160.

90. Pinker, *Language, Cognition, and Human Nature*, ix. Pinker's marketability is evident too in Penguin's repackaging and sale as short stand-alone books of *Hotheads* (2005), a chapter from *How the Mind Works* and *The Seven Words You Can't Say on Television* (2008), a chapter from *The Stuff of Thought*. Reviewing the work on swearing, one critic wrote: "Pulling out a chapter on rude words, giving it a striking and witty cover, may well be a shrewd move." Nicholas Lezard, "Stalking the Wild Taboos of Profanity," *Guardian*, October 4, 2008, 20.

91. Steven Pinker, *The Sense of Style: The Thinking Person's Guide to Writing in the 21st Century!* (New York: Penguin, 2014); and Steven Pinker, "Communicating Science and Technology in the 21st Century," accessed October 22, 2014, http://video.mit.edu/watch/communicating-science-and-technology-in-the-21st-century-steven-pinker-12644/.

CHAPTER FIVE

1. "Stephen Jay Gould: This View of Life," PBS *Nova*, producer Linda Harrar, aired December 18, 1984 (Paramus: NJ: Time-Life Video), VHS.

2. Louis P. Masur, "Stephen Jay Gould's Vision of History," *Massachusetts Review* 30 (1989): 482.

3. Stephen Jay Gould, *The Lying Stones of Marrakech* (New York: Harmony Books, 2000).

4. "Stephen Jay Gould: This View of Life."

5. Stephen Jay Gould, "The Streak of Streaks," *Triumph and Tragedy in Mudville: A Lifelong Passion for Baseball* (New York: W. W. Norton, 2003), 175.

6. Michael Shermer, "This View of Science: Stephen Jay Gould as Historian of Science and Scientific Historian, Popular Scientist and Scientific Popularizer," *Social Studies of Science* 32, no. 4 (2002): 489–24.

7. Stephen Jay Gould, *Eight Little Piggies: Reflections in Natural History* (New York: W. W. Norton, 1993).

8. Robin McKie, "DiMaggio of Science Hits a Last Home Run," *Observer*, May 7, 2000, 13.

9. Stephen Jay Gould, *The Mismeasure of Man* (London: Penguin, 1997), 53–54.

10. Natural History, "This View of Stephen Jay Gould," *Natural History* 108, no. 9 (1999): 48. The Library of Congress quote is taken from the jacket of Stephen Jay Gould, *The Structure of Evolutionary Theory* (Cambridge, MA: The Belknap Press of Harvard University Press, 2002).

11. Robert Wright, "The Accidental Creationist," *New Yorker*, December 13, 1999, 56.

12. Adam S. Wilkins, "Stephen Jay Gould (1941–2002): A Critical Appreciation," *BioEssays* 24, no. 9 (2002): 864.

13. Carol Kaesuk Yoon, "Stephen Jay Gould, 60, Is Dead; Enlivened Evolutionary Theory," *New York Times*, May 21, 2002, 5.

14. Stephen Jay Gould, *Rocks of Ages: Science and Religion in the Fullness of Life* (New York: Ballantine Publishing Group, 1999), 8.

15. Gould, *The Mismeasure of Man*.

16. Masur, "Stephen Jay Gould's Vision of History," 480.

17. Stephen Jay Gould, *The Structure of Evolutionary Theory* (Cambridge, MA: The Belknap Press of Harvard University Press, 2002), 38.

18. Myrna Perez, "Evolutionary Activism: Stephen Jay Gould, the New Left and Sociobiology," *Endeavor* 37, no. 2 (2013): 104–11.

19. Gould, *The Mismeasure of Man*.

20. Myrna Perez, "Evolutionary Activism," 106.

21. Stephen Jay Gould, *The Richness of Life: The Essential Stephen Jay Gould*, ed. Paul McGarr and Steven Rose (New York: W. W. Norton, 2007), 308.

22. Gould, *The Richness of Life*, 307.

23. John Horgan, "Escaping in a Cloud of Ink," *Scientific American,* August 1995, 40.

24. Stephen Jay Gould, *An Urchin in the Storm: Essays about Books and Ideas* (New York: W. W. Norton, 1987), 27.

25. Cited in "This View of Stephen Jay Gould," *Natural History* 108 (1999): 48.

26. Alvarez cited in Stephen Jay Gould, *Wonderful Life: The Burgess Shale and the Nature of History* (New York: W. W. Norton, 1989), 281.

27. H. Allen Orr, "The Descent of Gould," *New Yorker*, September 30, 2002, 132.

28. David F. Prindle, *Stephen Jay Gould and the Politics of Evolution* (Amherst, NY: Prometheus Books, 2009), 92.

29. Orr, "The Descent of Gould," 132, emphasis in original.

30. Perez, "Evolutionary Activism," 104–11.

31. Cited in "This View of Stephen Jay Gould," 48.

32. Cited in Perez, "Evolutionary Activism," 108.

33. Perez, "Evolutionary Activism," 108.

34. Stephen Jay Gould, *Full House: The Spread of Excellence from Plato to Darwin* (New York: Three Rivers Press, 1996), 8. The view is close to a Marxist philosophy of science that essentially argued science had unequaled power and record of achievement in understanding the natural world. But at the same time, it was "inextricably enmeshed with economic systems, technological developments, political movements, philosophical theories, cultural trends, ethical norms, ideological positions, indeed with all that was human." Helena Sheehan, "Marxism and Science Studies: A Sweep through the Decades," *International Studies in the Philosophy of Science* 21, no. 2 (2007): 197.

35. Elizabeth Allen et al., "Against Sociobiology," *New York Review of Books* 13 (1975): 182, 184–86.

36. Cited in Myrna Perez, "Evolutionary Activism," 108.

37. Myrna Perez Sheldon, "Claiming Darwin: Stephen Jay Gould in Contests over Evolutionary Orthodoxy and Public Perception, 1977–2002," *Studies in History and Philosophy of Biological and Biomedical Sciences* 45 (2014): 139–47.

38. David B. Wake, "Shape, Form, Development, Ecology, Genetics, and Evolution," *Paleobiology* 4, no. 1 (1978): 96.

39. It took three years, recalled the Norton editor Barber, once Gould had published more than thirty columns, that some of them could be published as a collection. Cited in "This View of Stephen Jay Gould," 48.

40. Orr, "The Descent of Gould," 132.

41. James Gorman, "The History of a Theory," *New York Times*, November 20, 1977, BR4.

42. Raymond A. Sokolov, "Talk with Stephen Jay Gould," *New York Times*, November 20, 1977, BR4.

43. Gould, *The Mismeasure of Man*, 52, 85–86.

44. Jim Miller, "What's in a Skull?" *Newsweek*, November 9, 1981, 106.

45. Gould, *The Mismeasure of Man*, 28.

46. Cited in Perez Sheldon, "Claiming Darwin," 140.

47. Gould, *An Urchin in the Storm*, 246.

48. Stephen Jay Gould, "Testimony of Dr. Stephen Jay Gould," *McLean v. Arkansas Documentation Project* (1981), accessed July 16, 2014, http://www.anti evolution.org/projects/mclean/new_site/depos/pf_gould_dep.htm.

49. Gould, *The Structure of Evolutionary Theory*, 989.

50. Cited in Barry Palevitz, "Love Him or Hate Him, Stephen Jay Gould Made a Difference," *Scientist*, June 10, 2002, 12.

51. For a history of creationism, see Ronald L. Numbers, *The Creationists: From Scientific Creationism to Intelligent Design* (Cambridge, MA: Harvard University Press, 2006).

52. Richard Dawkins, *The Blind Watchmaker* (London: Penguin, 1991), 251.

53. Wright, "The Accidental Creationist," 56.

54. Perez Sheldon, "Claiming Darwin," 140.

55. Jerry Adler and John Carey, "Enigmas of Evolution," *Newsweek*, March 29, 1982, 44.

56. Adler and Carey, "Enigmas of Evolution," 44.

57. Gould, *The Richness of Life*, 28–30.

58. John Durant, "In Memory of Stephen Jay Gould," *Public Understanding of Science* 11 (2002): 390.

59. "This View of Stephen Jay Gould," 48.

60. James Gleick, "Breaking Tradition with Darwin," *New York Times*, November 20, 1983, 48.

61. Spencer R. Weart, *Nuclear Fear: A History of Images* (Cambridge, MA: Harvard University Press, 1988), 382.

62. *The Climatic, Biological, and Strategic Effects of Nuclear War: Hearing Before the Subcommittee on Natural Resources, Agriculture Research, and Environment of the Committee on Science and Technology*, House of Representatives, 98th Cong., 8 (1985).

63. *The Climatic, Biological, and Strategic Effects of Nuclear War*, 8.

64. Lawrence Badash, *A Nuclear Winter's Tale: Science and Politics in the 1980s* (Cambridge, MA: MIT Press, 2009).

65. "Stephen Jay Gould: This View of Life."

66. Michelle Green, "Stephen Jay Gould," *People*, June 2, 1986, http://www.people.com/people/archive/article/0,,20093775,00.html.

67. Robert M. Ross, "Stephen Jay Gould: The Scientist as Educator," in *Stephen Jay Gould: Reflections on His View of Life*, ed. Warren D. Allmon, Patricia H. Kelley, and Robert M. Ross (New York: Oxford University Press, 2009), 245.

68. Warren D. Allmon, "The Structure of Gould: Happenstance, Humanism, History and the Unity of His View of Life," in *Stephen Jay Gould: Reflections On His View of Life*, ed. Warren D. Allmon, Patricia H. Kelley, and Robert M. Ross (New York: Oxford University Press, 2009), 43.

69. Stephen Jay Gould, *Wonderful Life: The Burgess Shale and the Nature of History* (New York: W. W. Norton, 1989), 52.

70. Gould, *Wonderful Life*, 14.

71. Gould, *Wonderful Life*, 277–81.

72. "This View of Stephen Jay Gould," 48.

73. David Papineau, "A Man with His Finger on the Future," *Independent*, January 31, 1993, 30.

74. Durant, "In Memory of Stephen Jay Gould," 390.

75. Pam Belluck, "Suit by Renowned Biologist's Widow Accuses Doctors of Negligence," *New York Times*, May 21, 2005, 14.

76. McKie, "DiMaggio of Science Hits a Last Home Run," 13.

77. Phillip Lopate, "Snails, Frankenstein and King Lear's Daughter," *New York Times*, January 21, 1996, 9.

78. Victoria Griffith, "The Man Who Calls Us Just 'an Accident,'" *Financial Times*, December 18, 1999, 3.

79. John Maynard Smith, "Genes, Memes and Mind," *New York Review of Books*, November 30, 1995.

80. Gould, *Full House*, 46–49, emphasis in original.

81. Suzanne Stephens, "Making it New in Soho," *Architectural Digest* 54, no. 2: 108–15.

82. Stephen Jay Gould, *Questioning the Millennium: A Rationalist's Guide to a Precisely Arbitrary Countdown* (New York: Harmony Books, 1999), 205.

83. Gould, *Rocks of Ages*, 9.

84. John Carey, "The Never-Ending Conflict," *Sunday Times*, January 28, 2001.

85. Mark Ridley, "The Evolution Revolution," *New York Times*, March 17, 2002, 11.

86. Michael Ruse, "The Gould Rush," *Globe and Mail*, March 23, 2002, D3.

87. Derek E. G. Briggs, "Stephen Jay Gould (1941–2002)," *Nature* 417, no. 6890 (2002): 706.

88. "Death of Author and Thinker," *New Scientist*, May 25, 2002, 9.

89. "Obituary of Stephen Jay Gould," *Daily Telegraph*, May 22, 2002, 23.

90. "Death of a Paleontologist," *New York Times*, May 21, 2002, 20.

91. As an indicator of cultural profile, I mapped the pattern of media attention to Gould, gathered from mentions of his name in the LexisNexis database, searched year by year under major world publication headings. This was not a formal content analysis, but was designed to map in some way patterns of attention to him over time. It should be noted that the archives from LexisNexis are often incomplete before 1980. See Anthony Downs, "Up and Down with Ecology: The 'Issue-Attention Cycle,'" *Public Interest* 28 (1972): 38–50.

92. "Obituary: Professor Stephen Jay Gould," *Independent*, May 22, 2002, 18.

93. Robin McKie, "A Grand Finale," *Observer*, May 26, 2002, 17.

94. Tim McLaughlin, "US Scientist Who Made Evolution Popular Dies," *National Post*, May 21, 2002, A2.

95. Wilkins, "Stephen Jay Gould (1941–2002)," 864.

96. Michiko Kakutani, "Professor in the Bleachers with a Lifelong Scorecard," *New York Times*, May 20, 2003, E7.

97. Clive Cookson, "An Inconclusive Experiment," *Financial Times*, May 24, 2003, 35.

98. Stephen Jay Gould, *The Hedgehog, the Fox, and the Magister's Pox* (New York: Harmony Books, 2003): 166.

99. Warren D. Allmon, Patricia H. Kelley, and Robert M. Ross, "Editor's Preface," in *Stephen Jay Gould: Reflections on His View of Life*, ed. Warren D. Allmon, Patricia H. Kelley, and Robert M. Ross (New York: Oxford University Press, 2009), ix.

100. Richard York and Brett Clark, *The Science and Humanism of Stephen Jay Gould* (New York: Monthly Review Press, 2011), 13.

101. Alexander Star, "Life's Work," *New York Times*, June 2, 2002, 18.

102. Richard C. Lewontin and Richard Levins, "Stephen Jay Gould—What Does It Mean to Be a Radical?" in *Stephen Jay Gould: Reflections on His View of Life*, ed. Warren D. Allmon, Patricia H. Kelley, and Robert M. Ross (New York: Oxford University Press, 2009), 204.

CHAPTER SIX

1. Royal Society of London, *The Public Understanding of Science* (London: Royal Society, 1985), 10.

2. Royal Institution, "RI History," *RI History*, accessed December 1, 2012, http://www.rigb.org/contentControl?action=displayContent&id=00000002894.

3. Lesley White, "Too Sexy for the Stuffed Shirts," *Sunday Times Magazine*, May 23, 2004, 20.

4. Cole Moreton, "The Girl with All the Brains," *Independent on Sunday*, May 11, 2008, 82.

5. Madeleine Kingsley, "Making Her Mark in the Male-Dominated World of Science," *Hello!* October 26, 1999, 106–9.

6. Sabine Durrant, "Don't Call Me My Dear," *Guardian*, May 31, 2000, 4.

7. Maev Kennedy, "People," *Guardian*, September 18, 2008, 23.

8. "Popularizer Greenfield Is Blackballed by Peers," *Nature* 429 (2004): 9.

9. John Bohannon, "The Baroness and the Brain," *Science* 310 (2005): 962–63.

10. Sean O'Hagan, "Desperately Psyching Susan: Sexy or Serious?" *Observer*, September 7, 2003, 5.

11. Stefanie Marsh, "All Work, No Play," *Times* (London), November 8, 2007, 10.

12. Steve Connor, "Controversial Professor Brands Campaign against Her Cowardly," *Independent*, February 7, 2004, 19.

13. Vanessa Thorpe, "Eureka! TV Scientists Turn Up a Fortune," *Observer*, April 18, 1999, 6.

14. Patricia Fara, "An Unlucky Fellow," *Times* (London), May 6, 2004, 14.

15. Gail Vines, "Susan Greenfield—a Mind of Her Own," *Independent*, September 20, 2003.

16. Tristram Hunt, "The Appliance of Science," *Independent*, November 20, 2001, 8.

17. Jane Kelly, "How the Daughter of a Chorus Girl Is Putting Sex into Science," *Daily Mail*, July 18, 2000, 11.

18. White, "Too Sexy for the Stuffed Shirts," 19.

19. Citation data from Web of Science database available through American University library website, http://subjectguides.library.american.edu/databasesatoz (accessed June 30, 2012).

20. Ted Anton, *Bold Science: Seven Scientists Who Are Changing Our World* (New York: W. H. Freeman, 2000).

21. Cited in Anton, *Bold Science*, 48.

22. Margaret Thatcher, "Speech to the Royal Society," September 27, 1988, accessed May 5, 2010, http://www.margaretthatcher.org/speeches/displaydocument.asp?docid=107346.

23. Cited in Anton, *Bold Science*, 41.

24. George H. W. Bush, *Presidential Proclamation 6158*, 1990, accessed November 5, 2012, http://www.loc.gov/loc/brain/proclaim.html.

25. David Jones, "Science Books about the Mind's Construction," *Times*, December 24, 1987.

26. For example, see: Susan Greenfield, "M-Words and the Brain," *Nature* 361 (1993): 127–28.

27. Nigel Williams, "Brain Story," *Current Biology*, October 1, 2000, accessed May 12, 2012, http://www.sciencedirect.com/science/article/pii/S0960982200007193.

28. Cited in Anton, *Bold Science*, 43.

29. Georgina Ferry, "Woman with Ideas on the Brain," *Independent*, December 27, 1994, 13.

30. Matthew Bond, "Pussy-Footing with the Bunny Girl Boss," *Times* (London), December 29, 1994.

31. Bond, "Pussy-Footing with the Bunny Girl Boss."

32. Nigel Hawkes, "Solving Ultimate Brain Teasers," *Times* (London), December 26, 1994, 26.

33. Ferry, "Woman with Ideas on the Brain,"13.

34. Orly Shachar, "Spotlighting Women Scientists in the Press: Tokenism in Science Journalism," *Public Understanding of Science* 9 (2000): 350.

35. Marcel Chotkowski LaFollette, *Science on American Television: A History* (Chicago: University of Chicago Press, 2013), 186.

36. Susan Greenfield, "Vision for Science," *Guardian*, November 14, 1998, 24.

37. Susan Greenfield, *The Human Brain: A Guided Tour* (London: Phoenix, 1998), xiv.

38. Greenfield, *The Human Brain*, xiv.

39. Greenfield, *The Human Brain*.

40. Susan Greenfield, "View from Here," *Independent*, January 29, 1998, E3.

41. John Cornwell, "It's a No Brainer," *Sunday Times*, April 27, 2008, 20–29.

42. White, "Too Sexy for the Stuffed Shirts," 21.

43. Decca Aitkenhead, "Brain Teaser," *Guardian*, June 8, 1998, 4.

44. Susan Greenfield, "We Need to Show the Inhabitants of the Modern Academic Monastery That There Is Another World Beyond," *Independent*, January 25, 2001; Susan Greenfield, "The Ivory Powerhouse," *Independent*, May 15, 1997, E3; Greenfield, "View from Here," E3.

45. PricewaterhouseCoopers, *The Future of Pharma HR*. PricewaterhouseCoopers, 2001, accessed May 12, 2012, www.pwc.fr/fr/pwc_pdf/pwc_the_future_of_p_hr.pdf.

46. Marsh, "All Work, No Play," 10.

47. Susan Greenfield, *2121: A Tale from the Next Century* (London: Head of Zeus, 2013), 392.

48. Ehsan Masood, "Highbrow 'Club' Seeks the Common Touch," *Nature* 396 (1998): 103.

49. Susan Greenfield, "Podium: Spreading the Gospel of Science," *Independent*, August 28, 1998, 4.

50. Tony Blair, *Science Matters*, 2002, accessed May 5, 2010, http://www.number-10.gov.uk/output/Page1715.asp.

51. O'Hagan, "Desperately Psyching Susan," 5. For a discussion of Blairism and media, see: Eugenio F. Biagini, "Ideology and the Making of New Labours," *International Labour and Working-Class History* 56 (1999): 93–105.

52. Sheila Jasanoff, *Designs on Nature: Science and Democracy in Europe and the United States* (Princeton, NJ: Princeton University Press, 2005).

53. Fiona Fox, "Practitioner's Perspective: The Role and Function of the Science Media Center," in *The Sciences' Media Connection: Public Communication and Its Repercussions*, ed. Simone Rödder, Martina Franzen, and Peter Weingart (New York: Springer, 2012).

54. Mwenya Chimba and Jenny Kitzinger, "Bimbo or Boffin? Women in Science: An Analysis of Media Representations and How Female Scientists Negotiate Cultural Contradictions," *Public Understanding of Science* 19, no. 5 (2010): 620.

55. Marsh, "All Work, No Play," 10.

56. Sarah Cassidy, "Greenfield: Create Centre to Aid Female Scientists," *Independent*, November 29, 2002, 10.

57. Elizabeth Leane, *Reading Popular Physics: Disciplinary Skirmishes and Textual Strategies* (Aldershot, UK: Ashgate, 2007), 160, emphasis in original.

58. Evelyn Fox Keller, *Reflections on Gender and Science* (New Haven, CT: Yale University Press, 1985).

59. Shachar, "Spotlighting Women Scientists in the Press."

60. Chimba and Kitzinger, "Bimbo or Boffin?" 622.

61. Chimba and Kitzinger, "Bimbo or Boffin?" 621.

62. Kingsley, "Making Her Mark in the Male-Dominated World of Science," 109.

63. Kingsley, "Making Her Mark in the Male-Dominated World of Science," 106.

64. Susan Greenfield, telephone interview with the author, March 17, 2014.

65. John Crace, "Dressed to Thrill," *Guardian*, February 8, 2000, 4.

66. Durrant, "Don't Call Me My Dear," 4.

67. Moreton, "The Girl with All the Brains," 30.

68. Peter Stanford, "How Do I Look? Lab Fab—Susan Greenfield," *Independent*, August 12, 2000, 6–7.

69. Robin McKie, "The Observer Profile: Susan Greenfield," *Observer*, July 9, 2000, 29.

70. Kelly, "How the Daughter of a Chorus Girl Is Putting Sex into Science," 11.

71. John Naish, "Chief of the Screen Police," *Times* (London), October 21, 2006, 6.

72. Bohannon, "The Baroness and the Brain," 962.

73. Susan Greenfield, telephone interview with the author, March 17, 2014.

74. Stanford, "How Do I Look? Lab Fab—Susan Greenfield,"6.

75. Marsh, "All Work, No Play," 10.

76. Sharon Churcher, "Peter's Brain Was a Real Aphrodisiac, But Now I Can Wear Short Skirts without Being Nagged," *Daily Mail*, May 4, 2003, 22.

77. White, "Too Sexy for the Stuffed Shirts."

78. Marsh, "All Work, No Play," 10.

79. Durrant, "Don't Call Me My Dear," 4.

80. Stanford, "How Do I Look? Lab Fab—Susan Greenfield," 6.

81. Sara Lawrence, "I Want a Bottom Like Kylie Minogue's," *Times* (London), July 24, 2003, 10. She later said it was a flippant quote, but was surprised the quote caused controversy, with scientists telling her it was an unwise thing to do, even though she asked: "Does it make me any less of a scientist because I think Kylie has a great bum? That's ludicrous" (cited in O'Hagan, "Desperately Psyching Susan," 5).

82. Stephen Cook, "UK . . . Walking: Greenfield Site," *Guardian*, August 12, 2000, 7.

83. Susan Greenfield, "Educated Eater," *Times* (London), October 12, 1996.

84. "Don't Ask Me How To: Speak French or Make Gravy," *Observer*, August 25, 2002, 1.

85. Marsh, "All Work, No Play," 10.

86. Fara, "An Unlucky Fellow," 14.

87. O'Hagan, "Desperately Psyching Susan," 5.

88. Helen Weathers, "Love, Sex and Our Marriage Split—by Britain's Brainiest Couple," *Daily Mail*, May 1, 2003, 46.

89. Neil Sears, "Sexual Chemistry Dies as Science's Most Glamorous Couple Separate," *Daily Mail*, April 30, 2003, 7.

90. Churcher, "Peter's Brain Was a Real Aphrodisiac," 22.

91. Ephraim Hardcastle, "Ephraim Hardcastle," *Daily Mail*, April 29, 2003, 13.

92. Weathers, "Love, Sex and Our Marriage Split," 46.

93. Cited in Weathers, "Love, Sex and Our Marriage Split," 47.

94. Churcher, "Peter's Brain Was a Real Aphrodisiac," 22.

95. Susan Greenfield, telephone interview with the author, March 17, 2014.

96. Fara, "An Unlucky Fellow," 14.

97. Chimba and Kitzinger, "Bimbo or Boffin?" 618.

98. Lisa Jardine, "The Many Faces of Science," *Nature* 405 (2000): 398.

99. Roy Porter, "Minds of Our Own," *Times* (London), June 14, 2000.

100. Susan Greenfield, *Tomorrow's People: How 21st-Century Technology Is Changing the Way We Think and Feel* (London: Penguin, 2004), ix–x.

101. Greenfield, *Tomorrow's People*, x.

102. Helen Zaltzman, "Books," *Observer*, December 5, 2004, 18.

103. Bohannon, "The Baroness and the Brain," 962.

104. Durrant, "Don't Call Me My Dear," 4.

105. Mark Henderson, "Is Susan Greenfield Too Famous to Be a Fellow?" *Times* (London), February 6, 2004, 1.

106. Cited in O'Hagan, "Desperately Psyching Susan," 5.

107. Cited in O'Hagan, "Desperately Psyching Susan," 5.

108. Tim Radford, "The Guardian Profile: Susan Greenfield." *Guardian*, April 30, 2004, 15.

109. Gillian Bowditch, "Inside the Mind of a Brain Expert," *Sunday Times*, February 26, 2006, 3.

110. Connor, "Controversial Professor Brands Campaign against Her Cowardly," 19.

111. Roger Highfield, "Science 'Fuddy-Duddies' Block Greenfield," *Daily Telegraph*, April 29, 2004, 4.

112. Connor, "Controversial Professor Brands Campaign Against Her Cowardly,"19.

113. Henderson, "Is Susan Greenfield Too Famous to Be a Fellow?" 1.

114. Cited in Connor, "Controversial Professor Brands Campaign against Her Cowardly," 19.

115. Vivienne Parry, "A Right Royal Rumpus," *Guardian*, February 12, 2004, 8.

116. Henderson, "Is Susan Greenfield Too Famous to Be a Fellow?" 1.

117. Roger Highfield, "Cowardly Whisperings Are Undermining Me, Says Woman Scientist," *Daily Telegraph*, September 22, 2003, 4.

118. Susan Greenfield, telephone interview with the author, March 17, 2014.

119. Susan Greenfield, *ID: The Quest for Meaning in the 21st Century* (London: Sceptre, 2009), 203.

120. Nigel Hawkes, "The Homogenisation of Society," The *Times* (London), May 17, 2008, 10.

121. Craig Brown, "It's All in Her Mind," *Mail on Sunday*, May 25, 2008, 11.

122. Cornwell, "It's a No Brainer," 23.

123. UK Parliament, *Parliamentary Business*, February 12, 2009: Column 1289, accessed November 5, 2012, http://www.publications.parliament.uk/pa/ld200809/ldhansrd/text/90212-0010.htm.

124. Susan Greenfield, "God Help Us All When Generation Text Are Running the Country," *Daily Mail*, August 12, 2009.

125. Susan Greenfield, "How Facebook Addiction Is Damaging Your Child's Brain." *Daily Mail*, April 23, 2009.

126. Susan Greenfield et al., "Modern Life Leads to More Depression among Children," *Daily Telegraph*, September 12, 2006, 23.

127. Catherine Bennett, "Baroness, You Are Being a Complete Twit about Twitter," *Observer,* March 1, 2009, 31.

128. Penny Wark, "Let's Screen Out Violence," *Times* (London), May 30, 2009, 5.

129. Erik Brown, "The Mind and the Media," *Mayfair Times*, April 2009, 20.

130. Greenfield planned to take legal action, based on sex discrimination. The matter was settled legally, with the proviso that neither side discuss it publicly.

131. "Royal Institution: Sparks Fly," *Guardian*, January 11, 2010, 32.

132. As an indicator of cultural profile, I mapped the pattern of media attention to Greenfield, gathered from mentions of her name in the LexisNexis database, searched year by year under major world publication headings. This was not a formal content analysis, but was designed to map in some way patterns of attention to her over time. It should be noted that the archives from LexisNexis are often incomplete before 1980. See Anthony Downs, "Up and Down with Ecology: The 'Issue-Attention Cycle,'" *Public Interest* 28 (1972): 38–50.

133. Frank Swain, "Susan Greenfield: Living Online Is Changing Our Brains," *New Scientist*, August 3, 2011, accessed July 16, 2012, http://www.newscientist .com/article/mg21128236.400-susan-greenfield-living-online-is-changing-our-brains.html#.U8hTIqj0uHk.

134. Dorothy Bishop, "An Open Letter to Baroness Susan Greenfield," *BishopBlog*, August 4, 2011, accessed May 20, 2012, http://deevybee.blogspot. com/2011/08/open-letter-to-baroness-susan.html.

135. T. McVeigh, "Research Linking Autism to Internet Use Is Criticised," *Observer*, August 6, 2011, accessed May 20, 2012, http://www.theguardian.com/society/2011/aug/06/research-autism-internet-susan-greenfield.

136. Anthony Giddens, *The Third Way: The Renewal of Social Democracy* (Cambridge: Polity Press, 1998).

137. Susan Greenfield, *You and Me: The Neuroscience of Identity* (London: Notting Hill Editions, 2011).

138. Susan Greenfield, "Attention, Please," *Literary Review* (Online), accessed November 7, 2012, http://www.literaryreview.co.uk/greenfield_09_10.html.

139. Greenfield, *You and Me*, 129.

140. Ben Goldacre, "Why Won't Professor Susan Greenfield Publish This Theory in a Scientific Journal?" *Bad Science* (blog), November 3, 2011, accessed May 20, 2012, http://www.badscience.net/2011/11/why-wont-professor-greenfield-publish-this-theory-in-a-scientific-journal.

141. Susan Greenfield, telephone interview with the author, March 17, 2014.

142. Adam Roberts, "Future Imperfect," *Guardian*, July 13, 2013, 11.

143. Adam Rutherford, "I Can't See a Future for This Dystopia," *Observer*, July 21, 2013, 38.

144. Clive Cookson, "What's on Susan Greenfield's Mind?" *Financial Times*, May 25, 2013, 56.

145. Susan Greenfield, telephone interview with the author, March 17, 2014. For a full discussion of her ideas on digital culture, see Susan Greenfield, *Mind Change: How Digital Technologies Are Leaving Their Mark on Our Brains* (London: Rider, 2014).

CHAPTER SEVEN

1. James Lovelock, *Homage to Gaia: The Life of an Independent Scientist* (Oxford: Oxford University Press, 2001), 191.

2. Michael McCarthy, "Guru Who Tuned into Gaia and Gave First Climate Change Warning," *Independent*, May 24, 2004, 6–7.

3. Ian Irvine, "The Saturday Profile: James Lovelock—The Green Man," *Independent*, December 3, 2005, 44.

4. John MacLeod, "More Nuclear Energy, Not More Hot Air," *Daily Mail*, April 6, 2006, 15.

5. "Hot Potato," *Sunday Telegraph*, May 30, 2004, 23.

6. Hans J. Schellnhuber, "'Earth System' Analysis and the Second Copernican Revolution," *Nature* 402 (1999): C19–C23; Ian Irvine, "The Saturday Profile: James Lovelock—The Green Man," *Independent*, December 3, 2005, 44.

7. Lovelock, *Homage to Gaia*, 188.

8. John Gray, "James Lovelock: A Man for All Seasons," *New Statesman*, March 25, 2013.

9. Lovelock, *Homage to Gaia*, 253.

10. James Lovelock, *The Revenge of Gaia: Why the Earth Is Fighting Back—and How We Can Still Save Humanity* (London: Penguin, 2007), 188.

11. James Lovelock, *Gaia: A New Look at Life on Earth* (Oxford: Oxford University Press, 1995).

12. Lovelock, *Gaia*, 10.

13. Jon Turney, *Lovelock and Gaia: Signs of Life* (Cambridge: Icon Books, 2003), 26.

14. Turney, *Lovelock and Gaia*.

15. Lovelock, *Homage to Gaia*, 295.

16. Lawrence E. Joseph, *Gaia: The Growth of an Idea* (London: Arkana, 1991), 48.

17. John Ryle, "The Secret of Everything," *Independent*, September 22, 1991, 3.

18. James W. Dearing, "Newspaper Coverage of Maverick Science: Creating Controversy Through Balancing," *Public Understanding of Science* 4 (1995): 341–61.

19. James Lovelock and Sidney Epton, "The Quest for Gaia," *New Scientist*, February 5, 1975, 304–6.

20. Declan Fahy, "Science Magazines," in *Encyclopedia of Science and Technology Communication*, ed. Susanna Hornig Priest (New York: Sage, 2010), 727–31.

21. Lovelock and Epton, "The Quest for Gaia," 305.

22. Martin Sherwood, "Inventing Pandora's Box," *New Scientist*, February 6, 1975, 307–9.

23. Sherwood, "Inventing Pandora's Box," 309.

24. Kenneth L. Woodward and Lorraine Kisly, "Mother Earth," *Newsweek*, March 10, 1975, 49.

25. Turney, *Lovelock and Gaia*.

26. Lovelock, *Gaia*, viii.

27. Lovelock, *Homage to Gaia*, 309.

28. Lovelock, *Gaia*, vii.

29. René Dubos, "Gaia and Creative Evolution," *Nature* 282, no. 5735 (1979): 154–55.

30. Nicholas Russell, "The Importance of Being Respectable," *Independent*, October 3, 1988, 15.

31. Michael Ruse, *The Gaia Hypothesis: Science on a Pagan Planet* (Chicago: University of Chicago Press, 2013).

32. Doolittle cited in Turney, *Lovelock and Gaia*, 68.

33. Richard Dawkins, *The Extended Phenotype: The Long Reach of the Gene* (Oxford: Oxford University Press, 1999), 235.

34. Cited in Charles Mann, "Lynn Margulis: Science's Unruly Earth Mother," *Science* 252, no. 5004 (1991): 378–81.

35. Lovelock, *Homage to Gaia*, 264.

36. Cited in "James Lovelock," *Beautiful Minds*, BBC 4, first aired April 14, 2010.

37. Joseph, *Gaia: The Growth of an Idea*, 67.

38. Ruse, *The Gaia Hypothesis*, 204.

39. Carl Sagan, "The Terraformers Are Coming," *New York Times*, January 6, 1985, 6.

40. Tom Ferrell, "A Planetary Air-Conditioner," *New York Times*, January 6, 1985, Books, 6.

41. Lawrence E. Joseph, "Britain's Whole Earth Guru," *New York Times*, November 23, 1986, 67.

42. Stephen Bocking, *Nature's Experts: Science, Politics and the Environment* (New Brunswick, NJ: Rutgers University Press, 2006), 55–56.

43. Anna Bramwell, *Ecology in the 20th Century: A History* (London: Yale University Press, 1989).

44. David Lindley, "Is the Earth Alive or Dead?" *Nature* 332 (1988): 483–84.

45. Richard A. Kerr, "No Longer Willful, Gaia Becomes Respectable," *Science* 240, no. 4851 (1988): 393.

46. John Postgate, "Gaia Gets Too Big for Her Boots," *New Scientist*, April 7, 1988, 60.

47. Although an acknowledged simple model, Daisyworld aimed to show that, through feedback and self-regulation, a stable temperature for life was maintained on Earth. For more on Daisyworld, see James Lovelock, *The Ages of Gaia: A Biography of Our Living Earth* (Oxford, Oxford University Press, 2000), 41–62. Claims about Daisyworld's robustness as a model have varied considerably. For more, see Stephen H Schneider, "A Goddess of Earth or the Imagination of a Man," *Science* 291, no. 5510 (2001): 1906–7

48. Lovelock, *The Ages of Gaia.*

49. Lovelock, *The Ages of Gaia*, 194–99.

50. Lovelock, *The Ages of Gaia*, 204.

51. Dennis Flanagan, "His Big Idea," *New York Times*, September 25, 1988, 13.

52. Spencer R. Weart, *The Discovery of Global Warming* (Cambridge, MA: Harvard University Press, 2008).

53. Walter Schwarz, "Building a Church for Gaia," *Guardian*, September 5, 1992, 25.

54. Ryle, "The Secret of Everything," 3.

55. Ryle, "The Secret of Everything," 3.

56. Fred Pearce, "Dr. Lovelock's New Bedside Manner," *New Scientist* 132 (1991), 43.

57. Fred Pearce, "Gaia, Gaia: Don't Go Away," *New Scientist* 142 (1994).

58. John Lawton, "Earth System Science," *Science* 292, no. 5524 (2001): 1965.

59. The full citation to the Amsterdam Declaration can be found at: http://www.igbp.net/about/history/2001amsterdamdeclarationonearthsystemscience.4.1b8ae2 0512db692f2a680001312.html.

60. Mary Midgley, *Gaia: The Next Big Idea* (London: Demos, 2001), 13.

61. Freeman Dyson, *From Eros to Gaia* (London: Penguin, 1993).

62. John Gray, *Straw Dogs: Thoughts on Humans and Other Animals* (London: Granta, 2002), 33–34.

63. Lovelock, *Homage to Gaia*, 267.

64. Schneider, "A Goddess of Earth or the Imagination of Man," 1906.

65. Adolfo Olea-Franco, *"Homage to Gaia: The Life of an Independent Scientist by James Lovelock," Journal of the History of Biology* 35, no. 3 (2002), 601.

66. Irvine, "The Saturday Profile: James Lovelock," 44–45.

67. James Lovelock, "Nuclear Power Is the Only Green Solution," *Independent*, May 24, 2004, 31.

68. Michael McCarthy, "Only Nuclear Power Can Halt Global Warming," *Independent*, May 24, 2004, 1.

69. Lovelock, *Homage to Gaia*, 339.

70. Irvine, "The Saturday Profile: James Lovelock," 44–45.

71. The Intergovernmental Panel on Climate Change (IPCC) is the scientific-political body composed of representatives of the world's governments that examines climate change. For a full description, see: Spencer R. Weart, *The Discovery of Global Warming* (Cambridge, MA: Harvard University Press, 2008).

72. Lovelock, *The Revenge of Gaia*.

73. James Lovelock, "The Earth Is About to Catch a Morbid Fever That May Last as Long as 100,000 Years," *Independent*, January 16, 2006, 31.

74. James Lovelock, "Nuclear Energy Will Help Save Our Planet," *Sun*, July 12, 2006.

75. Karen Bickerstaff, Irene Lorenzoni, Nick Pidgeon, Wouter Poortinga, and Peter Simmons, "Reframing Nuclear Power in the UK Energy Debate: Nuclear Power, Climate Change Mitigation and Radioactive Waste," *Public Understanding of Science* 17, no. 2 (2008): 145–69.

76. As an indicator of cultural profile, I mapped the pattern of media attention to Lovelock, gathered from mentions of his name in the LexisNexis database, searched year by year under major world publication headings. This was not a formal content analysis, but was designed to map in some way patterns of attention to him over time. It should be noted that the archives from LexisNexis are often incomplete before 1980. See Anthony Downs, "Up and Down with Ecology: The 'Issue-Attention Cycle,'" *Public Interest* 28 (1972): 38–50.

77. Cited in Neil Carter, "The Politics of Climate Change in the UK," *WIREs Climate Change* (2014), doi:10.1002/wcc.274.

78. Tim Flannery, "A Great Jump to Disaster?" *New York Review of Books*, November 19, 2009.

79. Bill McKibben, "How Close to Catastrophe?" *New York Review of Books*, November 16, 2006.

80. Tim Flannery, "Climate Change Personified," *Washington Post*, August 22, 2006, C03.

81. Mike Hulme, *Why We Disagree about Climate Change: Understanding Controversy, Inaction and Opportunity* (Cambridge: Cambridge University Press, 2009), 14.

82. Jeff Goodell, "The Prophet of Climate Change: James Lovelock," *Rolling Stone*, November 1, 2007.

83. Decca Aitkenhead, "Enjoy Life While You Can," *Guardian*, March 1, 2008, 33.

84. Steve Connor, "The IOS Interview: James Lovelock, Environmental Scientist," *Independent on Sunday*, February 5, 2006, 27.

85. Boris Johnson, "We've Lost Our Fear of Hellfire, but Put Climate Change in Its Place," *Daily Telegraph*, February 2, 2006, 18.

86. Michael Hanlon, "A Scorched Earth? Let's Stay Cool," *Daily Mail*, January 31, 2006, 12.

87. Peter Forbes, "Jim'll Fix It," *Guardian*, February 21, 2009, Review, 8.

88. Nigel Farndale, "If We're All Doomed, Enjoy It While You Can," *Sunday Telegraph*, March 15, 2009, 19.

89. Susan Elderkin, "The Earth Will Take Only So Much Abuse Before It Strikes Back," *Sunday Telegraph*, January 29, 2006, 51. See, for example: Weart, *Nuclear Fear*, Peter Weingart et al., "Of Power Maniacs and Unethical Geniuses: Science and Scientists in Fiction Film," *Public Understanding of Science* 12 (2003): 227–28.

90. William Underhill, "Here's Dr. Doom," *Newsweek*, April 24, 2006.

91. John Gribbin, "Not Worried about Global Warming?" *Independent*, July 31, 2006, 20.

92. Robin McKie, "Who Will Save the Earth?" *Observer*, November 12, 2006, 22.

93. James Wood, *The Fun Stuff and Other Essays* (New York: Farrar, Straus and Giroux, 2012), 52.

94. Lee R. Kump, "A Second Opinion for Our Planet," *Science* 325, no. 5940 (2009), 539.

95. James Lovelock, *The Vanishing Face of Gaia: A Final Warning* (London: Allen Lane, 2009), 23, 43.

96. Tim Flannery, "A Great Jump to Disaster?"

97. Andrew Watson, "Final Warning from a Sceptical Prophet," *Nature* 458, no. 7241 (2009): 970.

98. John Gribbin and Mary Gribbin, *He Knew He Was Right: The Irrepressible Life of James Lovelock and Gaia* (London: Allan Lane, 2009). The book was published in the United States as *James Lovelock: In Search of Gaia*.

99. Roger Highfield, "Irrepressible Lovelock's Final Warning on Global Warming," *Daily Telegraph*, March 21, 2009, Books, 22.

100. Leo Hickman, "James Lovelock: Fudging Data Is a Sin against Science," *Guardian*, March 30, 2010, 10.

101. Ian Johnston, "'Gaia' Scientist James Lovelock: I Was 'Alarmist' about Climate Change," NBC News, April 23, 2012, accessed July 16, 2014, http://worldnews.nbcnews.com/_news/2012/04/23/11144098-gaia-scientist-james-lovelock-i-was-alarmist-about-climate-change.

102. James Delingpole, "'Only Global Fascist Tyranny Can Save Us Now' Says Nice Old Man," telegraph.co.uk, March 30, 2010.

103. Toby Tyrrell, "Gaia: The Verdict Is . . . ," *New Scientist*, October 26, 2013, 31. See also: Toby Tyrrell, *On Gaia: A Critical Investigation of the Relationship between Life and Earth* (Princeton, NJ: Princeton University Press, 2013).

104. "Gaia: The Death of a Beautiful Idea," *New Scientist*, October 26, 2013, 5.

105. William H. Schlesinger, "Requiem for a Grand Theory," *Nature Climate Change 3*, no. 679 (2013): 697.

106. James Lovelock, *A Rough Ride to the Future* (London: Allen Lane, 2014), 15.

107. Lovelock, *A Rough Ride to the Future*, 18.

108. Michael Bond, "The Living Heart of Things," *New Scientist*, August 24, 2013, 48.

CHAPTER EIGHT

1. "Desktop Diaries: Brian Greene," *Science Friday*, Public Radio International, October 28, 2011, accessed July 16, 2014, http://www.sciencefriday.com/about/desktop-diaries.html

2. John Yaukey, "Physicist Wraps Universe in String," *USA Today*, December 14, 1999, 10D.

3. Janet Maslin, "The Almost Inconceivable, But Don't Be Intimidated," *New York Times*, February 26, 2004, E7.

4. Andrew Grant, "The Man Who Plucks All the Strings," *Discover*, March 9, 2010, accessed August 1, 2012, http://discovermagazine.com/2010/extreme-universe/08-discover-interview-man-who-plucks-all-the-strings.

5. John H. Schwarz, "Review of *The Elegant Universe: Superstrings, Hidden Dimensions, and the Quest for the Ultimate Theory*," *American Journal of Physics* 68 (2000): 199–200.

6. John Schwarz, cited in Sean Miller, *Strung Together: The Cultural Currency of String Theory as a Scientific Imaginary* (Ann Arbor: University of Michigan Press, 2013), 213.

7. Helge Kragh, *Quantum Generations: A History of Physics in the Twentieth Century* (Princeton, NJ: Princeton University Press, 2002), 417–18.

8. Miller, *Strung Together*.

9. Brian Greene, *The Elegant Universe: Superstrings, Hidden Dimensions, and the Quest for the Ultimate Theory* (New York: Vintage, 2003), 278.

10. Paul Ginsparg and Sheldon Glashow, "Desperately Seeking Superstrings," *Physics Today*, May 1986, 7.

11. Sheldon Glashow, "Tangled in Superstring," *The Sciences* 28, no. 3 (1988): 28.

12. Cited in Kragh, *Quantum Generations*, 419.

13. John Horgan, *The End of Science: Facing the Limits of Knowledge in the Twilight of the Scientific Age* (New York: Abacus, 1998).

14. Lee Smolin, *The Trouble with Physics: The Rise of String Theory, the Fall of a Science, and What Comes Next* (New York: Houghton Mifflin, 2006).

15. Peter Galison, "Theory Bound and Unbound: Superstrings and Experiment," in *Laws of Nature: Essays on the Philosophic, Scientific, and Historical Dimensions*, ed. Friedel Weinert (Berlin: Walter de Gruyter, 1995): 369–408.

16. Stephen Hawking and Leonard Mlodinow, *The Grand Design* (New York: Bantam, 2010), 143.

17. Smolin, *The Trouble with Physics*, xvii.

18. Greene, *The Elegant Universe*, 5–15.

19. Miller, *Strung Together*, 5

20. George Johnson, "Space-Time: The Final Frontier," *New York Times*, February 21, 1999, 6.

21. Marcia Bartusiak, "With Strings Attached," *Washington Post*, March 7, 1999, X09.

22. Robin McKie, "Mind Your Peas . . . ," *Observer*, June 17, 2000, 12.

23. Laurie M. Brown, "Brian Greene: *The Elegant Universe: Superstrings, Hidden Dimensions, and the Quest for the Ultimate Theory*," *Isis* 95, no. 2 (2004): 327.

24. Rachel Edford, "The Elegance of *The Elegant Universe*: Unity, Beauty, and Harmony in Brian Greene's Popularization of Superstring Theory," *Public Understanding of Science* 16 (2007): 449.

25. Steven Weinberg, *Dreams of a Final Theory* (New York: Pantheon Books, 1992).

26. Greene, *The Elegant Universe*, 166–67.

27. John Maddox, "A Further String to the Believers' Bow," *Nature* 398 (1999): 766–67.

28. Edford, "The Elegance of *The Elegant Universe*."

29. Schwarz, "Review of *The Elegant Universe*," 199.

30. Schwarz, "Review of *The Elegant Universe*," 199–200.

31. Jennifer Senior, "He's Got the World on a String," *New York*, February 1, 1999, accessed August 1, 2012, http://nymag.com/nymetro/news/people/features/1011/.

32. Tom Rhodes and Steve Farrar, "In Search of the Theory of Everything," *Australian*, April 27, 1999, 13.

33. Shira J. Boss, "Physicist Strings Together Theory of Universe," *Christian Science Monitor*, July 22, 1999, accessed August 1, 2012, http://www.csmonitor.com/1999/0722/p16s1.html.

34. Yaukey, "Physicist Wraps Universe in String," 10D.

35. Alden M. Hayashi, "A Greene Universe," *Scientific American* 282 (2000): 36–40.

36. Claudia Kalb, "Physics Envy," *Newsweek*, July 31, 2000, 62.

37. Roger Ebert, "*Frequency*," Roger Ebert.com, April 28, 2000, accessed August 1, 2012, http://www.rogerebert.com/reviews/frequency-2000.

38. David A. Kirby, *Lab Coats in Hollywood: Science, Scientists, and Cinema* (Boston: MIT Press, 2011).

39. Dan Odenwald, "*Nova* Strings Together a Theory That's Got It All," *Current*, July 14, 2003, accessed August 1, 2012, http://www.current.org/wp-content/themes/current/archive-site/doc/doc0313string.html.

40. Jose Van Dijck, "Picturizing Science: The Science Documentary as Multimedia Spectacle," *International Journal of Cultural Studies* 9, no. 1 (2006): 5–24.

41. Dan Vergano, "'Universe' Artfully Juggles Theories," *USA Today*, October 28, 2003, 3D.

42. Helen Stewart, "The Theory of Everything," *Sunday Times*, November 2, 2003, 59.

43. Virginia Heffernan, "A Three-Part Excursion to the Universe of String, Where the Cool Scientists Are," *New York Times*, October 28, 2003, E5.

44. Anjana Ahuja, "He's Got an Answer for Everything," *Times* (London), February 19, 2004, 10.

45. Imre Karacs and Steve Connor, "Forget Posh, the Big Book Money Is in Physics," *Independent*, October 21, 2000, 5.

46. Brian Greene, *The Fabric of the Cosmos: Space, Time, and the Texture of Reality* (New York: Vintage Books, 2005), ix–19.

47. Meir Ronnen, "Physics for the Everyman," *Jerusalem Post*, August 6, 2004, 23.

48. Maslin, "The Almost Inconceivable, But Don't Be Intimidated," E7.

49. Joel Achenbach, "If You Believe There's No Time Like the Present, You'd Be Right. Or Would You Be Wrong? Better Ask Brian Greene. Whatever," *Washington Post*, March 11, 2004, C1.

50. Rob Crilly, "It's Quantum Physics?" *Herald*, March 13, 2004, 6.

51. Arminta Wallace, "Life, the Universe and Everything," *Irish Times*, March 9, 2004, 15.

52. Stephen Strauss, "Bart Simpson, the Universe and You," *Globe and Mail*, April 12, 2004, R3.

53. Freeman Dyson, "The World on a String," *New York Review of Books*, May 13, 2004.

54. Roger G. Newton, "Weird Science," *New York Times*, April 11, 2004, 12.

55. Peter Woit, *Not Even Wrong: The Failure of String Theory and the Search for Unity in Physical Law* (New York: Perseus, 2006).

56. Peter Woit, "Macroscope: Is String Theory Even Wrong?" *American Scientist* 90 (2002): 110–12.

57. Lee Smolin, "Unraveling Space and Time," *American Scientist* 92, no. 4 (2004): 371.

58. Joseph Polchinski, "All Strung Out?" *American Scientist* 95 (2007): 72–75.

59. Richard Dawid, "On the Conflicting Assessments of the Current Status of String Theory," *Philosophy of Science* 76 (2009): 984–96.

60. "The Herb Garden Germination," *The Big Bang Theory*, CBS, first broadcast April 7, 2011.

61. *The Colbert Report*, Comedy Central, first broadcast May 27, 2008.

62. Brian Greene, "Q&A: Brian Greene on Music and String Theory," *Nature* 465 (2010): 426.

63. Brian Greene, "Put a Little Science in Your Life," *New York Times*, June 1, 2008, 14.

64. Andew Lawler, "The Big Apple Does Science," *Science* 320 (2008): 165.

65. Alan Childs, "The Many Dimensions of Brian Greene," *Times* (London), July 1, 2010, 36–41.

66. Brian Greene, *The Hidden Reality: Parallel Universes and the Deep Laws of the Cosmos* (New York: Alfred A. Knopf, 2011).

67. As an indicator of cultural profile, I mapped the pattern of media attention to Greene, gathered from mentions of his name in the LexisNexis database, searched year by year under major world publication headings. This was not a formal content analysis, but was designed to map in some way patterns of attention to him over time. It should be noted that the archives from LexisNexis are often incomplete before 1980. See Anthony Downs, "Up and Down with Ecology: The 'Issue-Attention Cycle,'" *Public Interest* 28 (1972): 38–50.

68. Janet Maslin, "Multiple-Universe Theory Made, Well, Easier," *New York Times*, January 27, 2011, C4.

69. George Ellis, "The Untestable Multiverse," *Nature* 469 (2011): 294–95.

70. Amanda Gefter, "Thoughts Racing Along Parallel Lines," *New Scientist*, February 5, 2011, 30.

71. Boaz Huss, "The New Age of Kabbalah," *Journal of Modern Jewish Studies* 6 (2007):107–25.

72. *Fresh Air*, NPR, February 11, 2005.

73. Anjana Ahuja, "He's Got an Answer for Everything," *Times* (London), February 19, 2004, 10.

74. "Welcome to the Multiverse," Daily Beast, May 21, 2012, accessed June 20, 2014, http://www.briangreene.org/bg_beast.html.

75. Amanda Schaffer, "Science as Metaphor: Where Does Brian Greene Stand in the Pantheon of Physicists?" *Slate*, July 6, 2004, accessed August 1, 2012, http://www.slate.com/articles/arts/culturebox/2004/07/science_as_metaphor.html.

CHAPTER NINE

1. John F. Kennedy, "Special Message to the Congress on Urgent National Needs," May 25, 1961, accessed July 1, 2012, http://www.jfklibrary.org/Research/Ready-Reference/JFK-Speeches/Special-Message-to-the-Congress-on-Urgent-National-Needs-May-25-1961.aspx.

2. Lynn Spigel, *Welcome to the Dreamhouse* (Durham, NC: Duke University Press, 2001).

3. Neil DeGrasse Tyson, *The Sky Is Not the Limit: Adventures of an Urban Astrophysicist* (Amherst, NY: Prometheus Books, 2004); and Tyson's comments in Claudia Dreifus, "Voices: 10/4/57," *New York Times*, September 25, 2007, 3.

4. Neil deGrasse Tyson, *Space Chronicles: Facing the Ultimate Frontier* (New York: W. W. Norton, 2012).

5. Dreifus, "Voices: 10/4/57," 3.

6. Spigel, *Welcome to the Dreamhouse*, 162.

7. Tyson, *Space Chronicles*, 67.

8. Susan Kruglinski and Marion Long, "The 10 Most Influential People in Science," *Discover*, November 26, 2008, accessed August 1, 2012, http://discovermagazine.com/2008/dec/26-the-10-most-influential-people-in-science/article_view?b_start:int=1&-C=.

9. David Segal, "Star Power: As an Astrophysicist, Neil deGrasse Tyson Is a Universal Expert," *Washington Post*, December 16, 2007, M01.

10. Carl Zimmer, "King of the Cosmos," *Playboy*, January–February 2012, accessed June 27, 2012, http://www.haydenplanetarium.org/tyson/read/2012/01/01/king-of-the-cosmos.

11. Charles Whitaker, "Super Stargazer: Neil de Grasse Tyson Is the Nation's Astronomical Authority," *Ebony*, August 2000, 58.

12. William M. Banks, *Black Intellectuals: Race and Responsibility in American Life* (New York: W. W. Norton, 1996).

13. Banks, *Black Intellectuals*.

14. Cornel West, "The Dilemma of the Black Intellectual," *Cultural Critique* 1 (1985): 110.

15. Tyson, *The Sky Is Not the Limit*.

16. Tyson, *The Sky Is Not the Limit*, 135.

17. Tyson, *The Sky Is Not the Limit*, 135.

18. Tyson, *The Sky Is Not the Limit*, 136.

19. Neil deGrasse Tyson, *Just Visiting This Planet* (New York: Main Street Books, 1998), xiii–xiv.

20. Tyson, *The Sky Is Not the Limit*, 136.

21. Tyson, *The Sky Is Not the Limit*, 137.

22. Tyson, *The Sky Is Not the Limit*, 137, 140.

23. Robert Boynton, "The New Intellectuals," *Atlantic*, March 1995, accessed March 12, 2012, http://www.robertboynton.com/articleDisplay.php?article_id=23.

24. For representations of black masculinity, see Ronald L. Jackson and Celnisha Dangerfield, "Defining Black Masculinity as Cultural Property: An Identity Negotiation Paradigm," in *Intercultural Communication: A Reader*, ed. Larry Samovar and Richard Porter (Belmont, CA: Wadsworth, 2002), 120–30.

25. Cited in Bruce Caines, *Our Common Ground: Portraits of Blacks Changing the Face of America* (New York: Crown Publishers, 1994), 39, emphasis in original.

26. Jeff Chang, *Can't Stop, Won't Stop: A History of the Hip-Hop Generation* (New York: St. Martin's Press, 2005), 17.

27. "I was relatively flexible for my size (six feet two, 190 pounds) having been a performing member of two dance companies while in college, and I was in pretty good shape, having wrestled varsity in NCAA Division I" (Tyson, *The Sky Is Not the Limit*, 52).

28. Neil deGrasse Tyson, "Ph.D. Convocation Address," Columbia University, New York City, May 14, 1991, accessed August 1, 2012, http://www.hayden-planetarium.org/tyson/read/1991/05/14/phd-convocation-address.

29. Cited in Caines, *Our Common Ground*, 39. Nevertheless, Tyson presented himself and other physicists as powerful counterexamples to prevailing negative portrayals of blacks. During the National Society of Black Physicists (NSBP) meeting in 1992, an event that coincided with the Los Angeles riots that followed the acquittal in court of four white police officers on trial for beating a black motorist, Rodney King, Tyson told how he replaced his initial talk with observations of the "NSBP's immeasurable significance to the perception of blacks by whites in America" (Tyson, *The Sky Is Not the Limit*, 135).

30. Banks, *Black Intellectuals*, 238.

31. Henry Louis Gates, Jr., "Bad Influence," *New Yorker*, March 7, 1994, 94; and cited in Boynton, "The New Intellectuals," 1995.

32. Caines, *Our Common Ground*, 39.

33. For more on Banneker, see Banks, *Black Intellectuals*, 248.

34. Stuart Hall, *Critical Dialogues in Cultural Studies* (New York: Routledge, 1996).

35. Tyson, *The Sky Is Not the Limit*, 140.

36. Steve Mirsky, "When the Sky Is Not the Limit," *Scientific American* 282, no. 2 (2000): 30. Also, the museum solicited recommendations from a member of the Princeton astrophysics department, where Tyson was then undertaking postdoctoral work. The physicist recommended Tyson.

37. For the magazine's aims, see: *Natural History*, Mission Statement, accessed March 28, 2012, http://www.naturalhistorymag.com/page/mission-

statement. Circulation varies from issue to issue; these figures are based on the magazine's media kit for 2009.

38. A cross-section of his articles can be viewed at Tyson's Hayden Planetarium homepage: http://www.haydenplanetarium.org/tyson/read.

39. Tyson, *Space Chronicles*, 259–60.

40. Hayden Planetarium, "Neil deGrasse Tyson Curriculum Vitae," accessed July 26, 2012, http://www.haydenplanetarium.org/tyson/curriculum-vitae.

41. Jan Hoffman, "Public Lives: Putting a Milestone in Cosmic Perspective," *New York Times*, December 31, 1999, 2.

42. Hoffman, "Public Lives," 2.

43. Whitaker, "Super Stargazer," 58–60.

44. Cited in Whitaker, "Super Stargazer," 62.

45. "Neil deGrasse Tyson: Sexiest Astrophysicist," *People*, November 13, 2000, accessed June 26, 2012, http://www.people.com/people/archive/article/ 0,,20132902,00.html.

46. Peter D. Meltzer, "Star Collector," *Wine Spectator*, May 31, 2000, 19–20.

47. Tyson, *The Sky Is Not The Limit*, 1–2.

48. Scott Gabriel Knowles, "Books in Brief," *New York Times*, August 1, 2004, 12.

49. Tyson, *The Sky Is Not the Limit*, 43.

50. Lisa R. Messeri, "The Problem with Pluto: Conflicting Cosmologies and the Classification of the Planets," *Social Studies of Science* 40, no. 2 (2010): 187–214.

51. Neil deGrasse Tyson, *The Pluto Files: The Rise and Fall of America's Favorite Planet* (New York: W. W. Norton, 2009), 65.

52. Kenneth Chang, "Pluto's Not a Planet? Only in New York," *New York Times*, January 22, 2001, 1.

53. Kenneth Chang, "How I (Ken Chang) Tormented Neil deGrasse Tyson," *New York Times*, January 8, 2009, accessed August 8, 2012, http://tierneylab.blogs. nytimes.com/2009/01/08/how-i-ken-chang-tormented-neil-degrasse-tyson.

54. Tyson, *The Pluto Files*, xii.

55. Cited in Tyson, *The Pluto Files*, 101.

56. Cited in Tyson, *The Pluto Files*, 103.

57. Tyson, *The Pluto Files*, 173.

58. Tyson, *The Pluto Files*, 135.

59. Messeri, "The Problem with Pluto," 208–9.

60. Kenneth Chang, telephone interview with the author, August 14, 2013.

61. Dan Vergano, telephone interview with the author, August 14, 2013.

62. Tyson, *Space Chronicles*, 224.

63. Chris Mooney, *The Republican War on Science* (New York: Basic Books, 2006), 237, viii.

64. Tyson, *The Sky Is Not the Limit*, 16.

65. For a further discussion of *Nova*'s history, see Marcel Chotkowski LaFollette, *Science on American Television: A History* (Chicago: University of Chicago Press, 2013), 121–35.

66. Ned Martel, "Mysteries of Life, Time and Space (and Green Slime)," *New York Times*, September 28, 2004, 5.

67. As an indicator of cultural profile, I mapped the pattern of media attention to Tyson, gathered from mentions of his name in the LexisNexis database, searched year by year under major world publication headings. This was not a formal content analysis, but was designed to map in some way patterns of attention to him over time. It should be noted that the archives from LexisNexis are often incomplete before 1980. See Anthony Downs, "Up and Down with Ecology: The 'Issue-Attention Cycle,'" *Public Interest* 28 (1972): 38–50.

68. Alex Strachan, "More Than a Science Show," *Montreal Gazette*, September 27, 2004, D5.

69. Cited in Scott Veale, "Newly Released," *New York Times*, December 1, 2004, 9.

70. Ned Martel, "Mysteries of Life, Time and Space [and Green Slime]," *New York Times*, September 28, 2004, E5.

71. "*Nova ScienceNow* Names Dr. Neil DeGrasse Tyson as New Host for Science Magazine Series in 2006," PBS News, January 14, 2006, accessed July 26, 2012, http://www.pbs.org/aboutpbs/news/20060114_novasciencenow.html.

72. Frances Bonner, *Personality Presenters: Television's Intermediaries with Viewers* (London: Ashgate, 2011), 3–4.

73. Alex Strachan, "Science Show Is Fast-Paced, Deep," *Montreal Gazette*, July 9, 2008, D6.

74. Tyson, cited in "*Nova ScienceNow* Names Dr. Neil DeGrasse Tyson."

75. Marisa Guthrie, "A Cry to Pass the Science Test," *Daily News*, November 21, 2006, 87.

76. Felicia R. Lee, "A Science Show Courts 'Blue-Collar Intellectuals,'" *New York Times*, October 3, 2006, 7.

77. Segal, "Star Power," M01.

78. Lisa deMoraes, "The 'Nova' Man is Bursting with Bright Ideas," *Washington Post*, January 9, 2009, C07.

79. *New York Times* best seller list from February 18, 2007. As an indicator of his star status, it shared space on the list alongside Thomas Friedman's *The World Is Flat* (no. 9), Richard Dawkins's *The God Delusion* (no. 8), *Culture Warrior* by Bill O'Reilly (no. 10), and Barack Obama's *The Audacity of Hope* (no. 1). See: Best Sellers, *New York Times*, February 18, 2007, Section 7, 26.

80. Michael D. Lemonick, "Neil deGrasse Tyson," *Time*, May 3, 2007, accessed August 1, 2012, http://content.time.com/time/specials/2007/time100/article/0,288 04,1595326_1595329_1616157,00.html.

81. Center for Inquiry, "Neil deGrasse Tyson—Communicating Science to the Public," Point of Inquiry Podcast, November 16, 2007, accessed August 1, 2013, http://www.pointofinquiry.org/neil_degrasse_tyson_communicating_science_to_the_public/.

82. National Academy of Sciences and Institute of Medicine, *Science, Evolution, and Creationism* (Washington, DC: National Academies Press, 2001), xiii.

83. Big Think, "Neil deGrasse Tyson: Atheist or Agnostic?" BigThink.com, April 8, 2012, accessed October 30, 2014, http://bigthink.com/think-tank/neil-degrasse-tyson-atheist-or-agnostic.

84. Neil deGrasse Tyson, *Death by Black Hole and Other Cosmic Quandaries* (New York: W. W. Norton, 2007), 334.

85. Chris Heller, "Neil deGrasse Tyson: How Space Exploration Can Make America Great Again," *Atlantic*, March 5, 2012, accessed October 8, 2012, http://www.theatlantic.com/technology/archive/2012/03/neil-degrasse-tyson-how-space-exploration-can-make-america-great-again/253989/.

86. Cited in Jesse Kornbluth, "If I Were President . . . ," *New York Times*, August 21, 2011, 12.

87. Lauren Feldman, Anthony Leiserowitz, and Edward Maibach, "The Science of Satire: *The Daily Show* and *The Colbert Report* as Sources of Public Attention to Science and the Environment," in Amarnath Amarasingam, ed., *The Stewart/Colbert Effect: Essays on the Real Impact of Fake News* (Jefferson, NC: McFarland, 2011), 25–46.

88. "Tyson on Colbert Preparation." *Politico*, March 14, 2010, accessed March 1, 2012, video available at http://www.politico.com/multimedia/video/tyson-on-colbert-preparation.html.

89. "The Apology Insufficiency," *The Big Bang Theory*, CBS, first broadcast November 4, 2010.

90. "Brain Storm," *Stargate Atlantis*, Sci-Fi Channel, November 21, 2008.

91. Sholly Fisch, "Star Light, Star Bright . . . ," *Superman Action Comics*, 2012, 14: 26–31. For the Superman comic, Tyson also identified an existing star that could serve as a real-life basis for Krypton's red sun, named in the comic books as Rao. He also supplied the celestial coordinates so amateur astronomers could find this real counterpart to Rao—the more prosaically named LHS-2520.

92. Jorge E. Hirsch, "An Index to Quantify an Individual's Scientific Research Output," *Proceedings of the National Academy of Sciences* 102, no. 46 (2005): 165–72.

93. National Science Foundation, Award Abstract: 0852400, *StarTalk Radio*, accessed October 8, 2012, http://www.nsf.gov/awardsearch/showAward.do?AwardNumber=0852400

94. "A Conversation with Nichelle Nichols," *StarTalk Radio*, Season 2, Episode 24, Curved Light Productions, July 11, 2011, accessed March 1, 2012, http://www.startalkradio.net/show/a-conversation-with-nichelle-nichols/.

95. Cited in "A Conversation with Whoopi Goldberg," *StarTalk Radio*, Season 2, Episode 39, Curved Light Productions, December 11, 2011, accessed March 1, 2012, http://www.startalkradio.net/show/a-conversation-with-whoopigoldberg/.

96. Rebecca Mead, "Starman," *New Yorker*, February 17, 2014, 80–87.

97. Roger Handberg, *Reinventing NASA: Human Spaceflight, Bureaucracy, and Politics* (Westport, CT: Praeger, 2003); and Mark E. Byrnes, *Politics and Space: Image Making by NASA* (Westport, CT: Praeger, 1994).

98. Freeman Dyson, "The 'Dramatic Picture' of Richard Feynman," *New York Review of Books*, July 14, 2011, accessed July 14, 2014, http://www.nybooks.com/articles/archives/2011/jul/14/dramatic-picture-richard-feynman/.

99. When NASA shaped its public image, it combined nationalistic, romantic, and pragmatic appeals. But it emphasized different appeals at various historical moments to match the prevailing public and political mood. It stressed nationalistic ideas during the Mercury era—from 1958 to 1963—as its funding increased amid fears of Soviet domination of space. Its nationalistic appeals continued in the Apollo era—from 1964 to 1972—but public disillusionment with the Vietnam War meant the agency also used romantic appeals about the heroic conquest of space. It changed its approach again in the shuttle era—from 1973 to 1990—as a dramatic drop in political support meant the agency had to stress the pragmatic results of space exploration, such as new technologies and products; see Byrnes, *Politics and Space,* 1994.

100. Neil deGrasse Tyson, "The Case for Space: Why We Should Keep Reaching for the Stars," *Foreign Affairs*, March–April 2012, accessed October 11, 2012, http://www.foreignaffairs.com/articles/137277/neil-degrasse-tyson/the-case-for-space.

101. Nick Zieminski, "Book Talk: The Case for a U.S. Mission to Mars, Reuters, March 8, 2012, accessed October 11, 2012, http://www.reuters.com/article/2012/03/08/us-books-authors-tyson-idUSBRE8270FC20120308.

102. *Late Night with Jimmy Fallon*, ABC, August 9, 2011, video accessed March 1, 2012, http://www.haydenplanetarium.org/tyson/watch/2011/08/09/late-night-with-jimmy-fallon-twitter-questions.

103. *The Colbert Report*, Comedy Central, April 8, 2010, video accessed March 1, 2012, http://www.haydenplanetarium.org/tyson/watch/2010/04/08/the-colbert-report.

104. Roger Pielke Jr., *The Honest Broker: Making Sense of Science in Policy and Politics* (Cambridge: Cambridge University Press, 2007).

105. "When Science Crashes the Party," *StarTalk Radio*, Season 2, Episode 38, Curved Light Productions, December 5, 2011, accessed March 1, 2012, http://www.startalkradio.net/show/when-science-crashes-the-party/.

106. Mead, "Starman."

107. John Heilpern, "Getting Astrophysical," *Vanity Fair*, May 2012, accessed July 12, 2014, http://www.vanityfair.com/hollywood/2012/05/neil-degrasse-tyson-snooki-jersey-shore.

108. The Science & Entertainment Exchange, "*Cosmos: A Spacetime Odyssey*," April 1, 2014, accessed July 1, 2014, http://www.scienceandentertainment exchange.org/article/cosmos-spacetime-odyssey.

109. James Rocchi, "Neil deGrasse Tyson, Ann Druyan, Seth MacFarlane Discuss '*Cosmos*,'" *Los Angeles Times*, June 12, 2014.

110. These figures are from ratings company Nielsen cited in Scott Collins, "Neil deGrasse Tyson's '*Cosmos*' Premiere Ratings: 40M First Week?" *Los Angeles Times*, March 10.

111. Hank Stuever, "Cosmos: A Fond Return to the Vastness of Space," *Washington Post*, March 9, 2014, T01.

112. Geoff Berkshire, "TV Review: '*Cosmos*,'" *Variety*, March 7, 2014.

113. Meredith Blake, "'*Cosmos*' Supporters Are Aiming to Make a Big Bang," *Los Angeles Times*, March 8, 2014, D1.

114. Dennis Overbye, "A Successor to Sagan Reboots '*Cosmos*,'" *New York Times*, March 4, 2014, 2.

115. Audra Wolfe, "Why *Cosmos* Can't Save Public Support for Science," *Atlantic*, March 11, 2014.

116. Charles C. W. Cooke, "Smarter than Thou—Neil deGrasse Tyson and America's Nerd Problem," *National Review*, July 21, 2014.

117. Jonathan H. Adler, "Does Neil deGrasse Tyson Make Up Stories?" *Washington Post*, September 22, 2014, accessed September 27, 2014, available from http://www.washingtonpost.com/news/volokh-conspiracy/wp/2014/09/22/does-neil-degrasse-tyson-make-up-stories/.

CHAPTER TEN

1. Stephen Jay Gould, "Bright Star among Billions," *Science* 275, no. 5300 (1997): 599.

2. David A. Kirby, *Lab Coats in Hollywood: Science, Scientists, and Cinema* (Cambridge, MA: MIT Press, 2011).

3. Louis Menand, "The Iron Law of Stardom," *New Yorker*, March 7, 1997, 36–39.

4. Simon Blackburn, "The Blank Slate," *New Republic*, November 25, 2002, 28.

5. Evelyn Fox Keller, *Reflections on Gender and Science* (New Haven, CT: Yale University Press, 1985), 5.

6. John Maynard Smith, "Genes, Memes and Mind," *New York Review of Books*, November 30, 1995.

7. J. G Ballard, *A User's Guide to the Millennium* (London: HarperCollins, 1996), 149.

8. John Lawton, "Earth System Science," *Science* 292, no. 5524 (2001): 1965.

9. Warren D. Allmon, "The Structure of Gould: Happenstance, Humanism, History and the Unity of His View of Life," in *Stephen Jay Gould: Reflections on His View of Life*, ed. Warren D. Allmon, Patricia H. Kelley, and Robert M. Ross (New York: Oxford University Press, 2008), 30.

10. Steven Pinker, *How the Mind Works* (New York: W. W. Norton, 1997), x.

11. Steven Pinker, *Language, Cognition, and Human Nature: Selected Articles* (New York: Oxford University Press, 2013), x.

12. Pinker, *Language, Cognition, and Human Nature*, x.

13. Louis Menand, "Dangers Within and Without," *Profession*, 2005, 10–17. As Louis Menand wrote, instead of lamenting this type of large-scale writing, humanities scholars should provide an alternative to it.

14. Matthew Kalman, "Hawking Backs Academic Boycott of Israel by Pulling Out of Conference," *Guardian*, May 8, 2013, 3; Press Association, "Hawking Speaks of Brush with Death in 1980s," *Guardian*, July 8, 2014, 8; Ian Sample, "A Brief History of Team," *Guardian*, May 29, 2014, 5.

15. Steven Pinker, telephone interview with the author, March 4, 2014.

16. "Professor Brian Cox OBE," Sue Rider Management, July 16, 2014, http://www.sueridermanagement.co.uk/presenters/BrianCox/briancox.htm.

17. Jane Fryer, "The Man Who's Making Space Sexy," *Daily Mail*, March 31, 2010.

18. Roger Highfield, "The Cox Effect Is a Star Turn," *Daily Telegraph*, September 6, 2011, 29.

19. Colin Tudge, "The Science Gurus," *Independent*, January 5, 1997. For a sociological examination of how a scientist's visibility affects their reputation and wider scientific work, see: Massimiano Bucchi, "Norms, Competition and Visibility in Contemporary Science: The Legacy of Robert K. Merton," *s* (2014), DOI: 10.1177/1468795X14558766.

Index

Morris, Errol, 27
Morton, Oliver, 73
Morton, Samuel George, 94–95
movies, 12, 152, 166–67, 170, 185,
    197. *See also* documentaries;
    individual films
M-theory, 35, 161
multiverse, 35, 172–74
*Music to Move the Stars: A Life with
    Stephen* (Jane Hawking), 29
My Brief History (Hawking), 38

Nabokov, Vladimir, 108
Nagel, Thomas, 54
napalm, 8
NASA, 179, 195–97
National Academy of Sciences, 5, 197
National Review, 199–200
National Science Foundation (NSF),
    194
National Society of Black Physicists
    (NSBP), 261n29
nationalism, 196
Natural History: Gould's columns,
    88–90, 92, 93, 98, 102–3, 208;
    Tyson's Universe column, 185,
    188, 192, 200
natural selection, 1, 44, 47, 137, 139–
    40, 207, 210
The Nature of the Universe (Hoyle), 3
nature vs. nurture, 76, 78
Nazis, 67, 84, 89
neo-Darwinism, 44, 45–46, 48, 59, 63
nerd culture, 199–200
neuroscience, 113; consciousness,
    nature of, 115–16, 124;
    popularizing, 115–17, 124–25
New York Times best seller list, 4, 25,
    28, 48, 60, 62, 79, 83, 192, 196
newspapers: science sections, 3–4, 10
Newton, Isaac, 23–24, 26

Nichols, Nichelle, 195
Nobel Prize, 6, 9, 111
non-overlapping magisteria (NOMA),
    105–6
Not Even Wrong (Woit), 169–70
Nova (television series), 4, 87, 100;
    "The Fabric of the Cosmos," 170,
    171, 174; "The Elegant Universe,"
    167–68; Tyson and, 180, 189–94
Nova ScienceNow, 191
nuclear power, 10; as alternative to
    global warming, 148–49, 151
nuclear winter, 99–100, 152
Nye, Bill, 206

Official View, 76
O'Hagan, Sean, 118, 122
Olea-Franco, Adolfo, 148
On Gaia (Tyrrell), 155
On the Shoulders of Giants (Hawking),
    32
Ontogeny and Phylogeny (Gould), 93
The Origin of Species (Darwin), 1–2,
    60, 92, 145
*Origins: Fourteen Billion Years of
    Cosmic Evolution* (Tyson), 191–92
Orr, H. Allen, 54, 91, 93
*Our Common Ground: Portraits of
    Blacks Changing the Face of America*
    (Caines), 184
Overbye, Dennis, 27, 167, 199
Overton, William, 96
Oxford University, 50, 113
Oxford University Press, 44–45, 139

Pais, Abraham, 2
paleontology, 14, 89, 90–91; Burgess
    Shale fossils, 101–2, 108; dinosaur
    extinction, 99, 102
The Panda's Thumb (Gould), 94
Papineau, David, 81

About the Author

Declan Fahy, a former newspaper reporter, is a member of the School of Communication faculty at American University. He holds a doctorate in communication from Dublin City University and his scholarship has been published in *Journalism*, *Journalism Studies*, *Nature Chemistry*, *Science Communication*, *BMC Medical Ethics*, *Irish Communications Review*, and other journals. Before joining academia, he worked for nearly a decade as a professional reporter and features writer for the *Longford Leader*, *Irish Daily Mirror*, and *The Irish Times*. His recent journalism appears in the *Columbia Journalism Review* and the *Scientist*. He lives in Washington, DC. *The New Celebrity Scientists: Out of the Lab and Into the Limelight* is his first book.